THE NATURE OF IRREVERSIBILITY

THE UNIVERSITY OF WESTERN ONTARIO
SERIES IN PHILOSOPHY OF SCIENCE

A SERIES OF BOOKS
IN PHILOSOPHY OF SCIENCE, METHODOLOGY,
EPISTEMOLOGY, LOGIC, HISTORY OF SCIENCE,
AND RELATED FIELDS

VOLUME 28

THE NATURE OF IRREVERSIBILITY

A Study of Its Dynamics and Physical Origins

by

HENRY B. HOLLINGER
Department of Chemistry, Rensselaer Polytechnic Institute

and

MICHAEL JOHN ZENZEN
Department of Philosophy, Rensselaer Polytechnic Institute

D. REIDEL PUBLISHING COMPANY

A MEMBER OF THE KLUWER ACADEMIC PUBLISHERS GROUP

DORDRECHT / BOSTON / LANCASTER / TOKYO

Library of Congress Cataloging in Publication Data

CIP

Hollinger, Henry B., 1933–
The nature of irreversibility.

(The University of Western Ontario series in philosophy of science ; v. 28)
Bibliography: p.
Includes indexes.
1. Irreversible processes. 2. Fluid dynamics. 3. Statistical mechanics.
I. Zenzen, Michael John, 1945– II. Title. III. Series.
QC174.17.176H65 1985 530.1 85-18280

ISBN-13: 978-94-010-8897-8 e-ISBN-13: 978-94-009-5430-4
DOI: 10.1007/978-94-009-5430-4

Published by D. Reidel Publishing Company,
P.O. Box 17, 3300 AA Dordrecht, Holland.

Sold and distributed in the U.S.A. and Canada
by Kluwer Academic Publishers,
190 Old Derby Street, Hingham, MA 02043, U.S.A.

In all other countries, sold and distributed
by Kluwer Academic Publishers Group,
P.O. Box 322, 3300 AH Dordrecht, Holland.

PREFACE

A dominant feature of our ordinary experience of the world is a sense of irreversible change: things lose form, people grow old, energy dissipates. On the other hand, a major conceptual scheme we use to describe the natural world, molecular dynamics, has reversibility at its core. The need to harmonize conceptual schemes and experience leads to several questions, one of which is the focus of this book. How does irreversibility at the macroscopic level emerge from the reversibility that prevails at the molecular level?

Attempts to explain the emergence have emphasized probability, and assigned different probabilities to the forward and reversed directions of processes so that one direction is far more probable than the other. The conclusion is promising, but the reasons for it have been obscure. In many cases the aim has been to find an explanation in the nature of probability itself. Reactions to that have been divided: some think the aim is justified while others think it is absurd.

Other accounts of irreversibility have appealed to averaging procedures used in probabilistic calculations, to stochastic flaws in the underlying mechanical model, to the large numbers of molecules involved, to jammed correlations, unseen perturbations, hidden variables, uncertainty principles, and cosmological factors. While acknowledging these attempts as important articulations of basic ideas in statistical mechanics and other theories, we feel that they do not meet the irreversibility paradox head-on. They do not explain the origin of irreversibility as it occurs in the common, everyday, natural behavior of observable materials. We think the natural macroscopic irreversibility can be understood only by examining the predicament of the systems involved. Irreversibility is not an intrinsic feature of any system, however large or small, quantized or not, fast or slow, whatever. Rather it is a consequence of a predicament which can be identified in terms of the ratio between the typical times for internal changes and the typical times between external changes. Any mechanical system can perform reversibly in some predicament. Any predicament can allow some systems to be reversible and force others to be irreversible.

The issue for molecular dynamics is to understand the patterns of behavior for isolated systems. In that predicament, the behavior is reversible. Then the effects of external changes can be examined to determine where and how they induce irreversibility. Probability and statistical mechanics can be useful as tools in these examinations provided they are applied only to reproducible and predictable phenomena, and are never substituted for the roles played by molecular dynamics.

What is needed and largely missing in the current theories about molecular dynamics is a science of "equilibrium plateaus" in the temporal variations of collective properties. It seems to us that once the phenomenon of equilibrium is understood in terms of molecular dynamics, the macroscopic appearance of irreversibility can be understood in terms of the frequency of forced withdrawals from young equilibria. We believe that the paradox of irreversibility can be resolved in a simple, logically clear, and aesthetically pleasing manner.

In the absence of a science of equilibrium in many particle systems, we must proceed on empirical knowledge. For observed fluids the distribution of plateau lengths seems to be such that we can generally assume a plateau length is longer than necessary to induce irreversible behavior. That is our assumption throughout this book. The assumption is supported by sketchy estimates in our literature and by what we have found in computer simulations of rarefied gases, which should have the shortest plateaus.

This book is intended primarily for philosophers and scientists who have a special interest in the interpretive and foundational problems associated with statistical mechanics and the dynamics of irreversibility. For those readers who want to see details about how irreversibility enters fluid dynamics and particle dynamics and how it is treated in statistical mechanics, we have included additional chapters which reflect material used to complement the course here in nonequilibrium statistical mechanics. In those chapters we have tried to provide a rationale for the mathematics and to provide enough detail to be nearly self-contained while at the same time aiming the discussion at questions about irreversibility.

The description here belies the effort that went into the study. Both of us had been independently interested in questions about temporal anisotropy for many years before we began to collaborate. And we argued for several more years before we decided to try to gather together what history and conclusions we could. To our surprise, things came together and many of our arguments became slight differences over details. We now want to communicate our conclusions to other people who ponder irreversibility, and beyond that, to engage new participants.

There may be those like Hollinger who start out believing that time is merely a parameter for Newtonian motion. Relativity is merely an alteration in the scheme and does not argue against the fact that Newton can still sit in his frame (maybe not even inertial) and expect to calculate everything that can happen in any frame, including slowed clocks, rotations, and whatever comes out of any new theories. And there may be others like Zenzen who have the physics background but resist the dogma and want to leave open certain questions about time in its most general and abstract aspect. The two of us have been bent toward each other's view, and we have found agreement on many details. Hollinger must now admit that time is a bigger mystery than he assumed in those years, and Zenzen agrees that Newtonian time, for all its simplicity, does allow for a wider range of explanation than is generally recognized. Neither feels that he has reached the end in these studies, but each feels that he has made some progress.

We are grateful to Professor Robert Butts and other colleagues who have offered encouragement and/or critical comments, all of which have affected the presentation. We are grateful to Geri Frank for her masterful editing and typing. We are grateful to the Philosophy of Science Association for permission to use parts of our article in Philosophy of Science, 49 (1982) pp.309-354.

TABLE OF CONTENTS

CHAPTER 1

INTRODUCTION

Some of the most obvious features of the world and our experience are the processes of dissipation and aging. Things break down, wear out, become less efficient and we generally have the sense that, unless something is done to rejuvenate them, systems will inevitably decay to the point where they will lose the very attributes that gave them identity as "systems."

But while we all recognize this pervasive characteristic of aging, we are also aware of the fact that this is a relative property of physical systems. Thus current cosmological theory informs us that the sun was born and will die, but with respect to our process of aging, the sun seems to be in quite a stable state. Everything may be in flux (one can even talk of the half life of a proton) but an equally acceptable phrasing is that there are degrees of stability.

On the one hand we have the notion of asymmetric processes during which systems age, and on the other, the idea of stasis where by physical systems seem to be resting, compared to more turbulent ones. These are undoubtedly rough and unsophisticated notions but they serve as large baskets into which we can begin putting chunks of experience. Thinking about the world in an untutored way, there is a *prima facie* plausibility about trying to sort things according to whether they seem fixed or in flux relative to us and whether those in flux seem to have something in common, such as a fundamental asymmetry.

If this sorting works, then it is reasonable to seek an explanation of how things change or how things remain (relatively) stable. It is a commonplace of Western intellectual history that the speculations of early Greek atomistics led to a form of explanation which tries to account for change in terms of the configurations and movements of the essentially changeless and "uncuttable." In recent years some philosophers and scientists have noted that contemporary science frequently makes use of an inverse form of explanation. Thus, rather than trying to account for change, the problem is to explain how systems achieve and maintain some degree of stability.

Of course, this way of describing things is very abstract. Actually the term "physical system" as we have been using it, covers living and nonliving things, liquids, solids, gases, plasmas, and crystals. Faced with this diversity it may seem quite presumptuous to look for some general pattern which might help us understand some of the relationships between stability and change. But we need not pretend that we are the first people to think about such issues. Indeed, much of the history of physics and physical chemistry can be seen as an on-going effort to sort out and describe dynamically various states, changes and fluxes of interesting and agreeable physical systems.

And the most casual look at these efforts shows that they rely on some model, a kind of ideal construct which hopefully captures the essential features of the real system or process. A good model allows us to use mathematics, manipulate symbols and then study the behavior of the system by interpreting or reading the relevant symbols. This sounds straightforward and generally it is. But sometimes the mathematicization of the model is problematic, or the interpretation of the mathematics after manipulation is unclear, or the predicted behavior seems at odds with the observed behavior of the system.

The cluster of problems with which this book is concerned center on the question: How can the anisotropic, directional macrophysical behavior of a system of molecules be reconciled with the reversibility of the underlying molecular model? The clarification of this question and construction of an answer will take us through the highways and by-ways of particle mechanics, thermodynamics, kinetic theory, and statistical mechanics. As we shall see, this question has a complex history; it took quite a while to germinate and continues to exercise theorists. What appears to be a simple question has fostered explanations of remarkable sophistication. Thus, we have appeals to "laws of large numbers, jammed correlations, unseen perturbations, procedures (e.g., coarse graining and time smoothing), stochastic flaws, cosmological origins," etc. One might suspect that this is a case of obscurans per obscurantis if it were not for the fact that the explanation of irreversibility seems to require extreme measures.

Our bias is toward simplicity. While recognizing the power and utility of mathematics, we are also aware that it can obscure important concepts. A case in point: In early and modern discussions of reversibility, the reversibility of equations (time-reversal invariance) has been confused with the reversibility of physical processes. By focussing on the equations, treating them as fully isomorphic to the physical processes, important concepts not represented in the formalism are missed.

This motif is crucial to our analysis. Our reading of the antecedent and consequent history of the problem of irreversibility reveals an initial struggle for clarification and attempts to articulate the relevant concepts, followed by many misunderstandings and reformulations. What chances there were initially to get at the heart of the problem were severely compromised as the discussion became increasingly tied to formalism and theorems. In our opinion, the explanation of irreversibility requires several meta-principles which are not themselves mathematical expressions or conditions on the applicability of such expressions. If we are correct about this, then it is clear that more mathematical sophistication will not lead one out of the labyrinth of irreversibility.

In a sense then, our study is a return to the original problem context, an attempt to do a kind of conceptual archeology. We want to remove the sediment of formalism and return to the bedrock of the model to see what possibilities it offers. By returning to the original problem context and tracing the subsequent discussions we can get clues about where and for what reasons people introduced what were, in our view, unnecessary complications. These clues help us to make important clarifications and distinctions, a sort of preliminary taxonomy which can then be used to sort extraneous material.

Our anti-formalist views and bias towards simplicity might indicate to the reader (especially the scientific reader) that he is about to be subjected to the arid philosophical discourse which is commonly condemned as purely semantic. This is not the case. Those who need and enjoy the security of mathematical expressions will find them here, but when they appear it is not gratuitous. We shall not simply use equations: they will be objects for analysis and reflection. We are as much interested in what they convey as what they might obscure. It is in this spirit that we assess contemporary developments in kinetic theory, statistical mechanics, and nonequilibrium thermodynamics. We want to provide the reader with a map of current research which at least shows the contours of major programs. This, of course, is not a neutral description and we try to evaluate these programs according to explicit criteria.

One of the difficulties which authors must face is what to say first, what to emphasize, what to put aside. We realized early in our collaboration that technical, analytical, historical, and philosophical considerations could occupy us endlessly. We simply had to decide to suspend discussion of many things so that the book had some continuity. The enormous amount of material on the family of topics which include irreversibility and our tendencies to follow our thoughts wherever they might lead made this a hard decision to live by. For most of the book we restrained ourselves but in the end we indulged in a set of philosophical reflections about time, consciousness, randomness, probability, etc.

CHAPTER 2

THE PARADOXES

Before we begin our investigation of irreversibility it is important to look at the historical development of the concept and its problems. Not only will this give us a historical orientation but it will help also to illustrate the difficulties involved in giving formal expression to intuitive and qualitative notions and shed light on the current situation.

The historical perspective which we provide here is not neutral or unbiased (if such be possible). There is a moral to the story we tell and it goes something like this: insight into fundamental problems of science may be enhanced by expressing the problem in formal language but this operation has both clarifying and obscuring aspects. The precision and clarity gained from mathematics is sometimes coupled with the sacrifice of concepts or principles which may be meta-mathematical or "physical" in a sense which does not allow a complete translation into formal terms. The theoretician who gets caught up in the demanding mathematics of a discipline (e.g., statistical mechanics) is not likely to recover such concepts or principles.

With this then in mind we begin our story. Sometimes it may seem a bit dry and long-winded to the reader but if you keep the "moral" in mind, it will make things easier. There will then be a dynamic to the history and the stage will be set for the reconstruction which forms the basis of our analysis in the next chapter (Note 1).

2.1 Early Studies of Heat and Attempts to Formulate Equations of Heat Flow

Heat flows from hot bodies to colder ones, never the other way. This simple observation is at the heart of thermodynamics, but, the second law is not simply a reformulation of this notion in mathematical terms. People noticed and theorized about heat flow long before the development of thermodynamics. What phenomena did they study and how did they use the concept of irreversibility?

Some of the earliest systematic studies of heat were concerned with how the earth was formed. Speculations that the core of the earth was molten were stimulated by such natural phenomena as geysers, volcanos, and igneous rock. It was also known that the temperature of the earth gradually increased as one moved toward the center. Various calculations were made and eventuated in a theory of gradual refrigeration; thus a "heat depth" of the universe was published as early as 1777 by J.S. Bailly in Lettres sur l'Origine des Sciences. This claim set the stage for a debate that was to extend into the 19th century, a debate which was central to geophysical theory. On one side were James Hutton and his disciple, John Playfair and on the other, Richard Kirwan, John Murray and John Hunter.

Hutton propounded a cyclic theory in which continents were destroyed and reformed; the driving mechanism for this is the central fires of the earth. Hunter and Murray challenged this with the observation that heat diffuses throughout a body until it reaches a uniform temperature; there cannot be any cycling of the sort that Hutton proposed.

For our purposes, the significant point about these debates is that they lead to attempts to formulate equations of heat flow. Biot (1804) recognized that a general law of heat conduction might be based on the assumption that the rate of flow is proportional to the temperature difference between the two bodies. The work of Joseph Fourier (1808) established the basis for the modern theory of heat conduction and clearly expressed the irreversible character of heat flow. A foreshadowing of later difficulties can be seen here. Fourier's equations were not symmetrical with respect to past and future directions of time as were those of Newtonian mechanics. Some scientists saw this as a basic contradiction but others had a more liberal interpretation of "Newtonian mechanics" and saw no reason why it should be restricted to systems which have no dissipation or velocity dependent forces. If these differences in viewpoint had been fully articulated, later confusions might have been avoided. But as it happened, Helmholtz's formulation of the principle of energy conservation in terms of forces between point particles and the revival of kinetic theory of gases by Clausius and Maxwell tended to reinforce the view of "Newtonian mechanics" as essentially and fundamentally time-reversible. As we shall see, this sharp separation of a reversible underlying mechanics from dissipative, irreversible phenomena is the recurring motif in the story of irreversibility.

2.2 The Problems of Heat Distribution in the Earth and Thomson's 1852 Statement on Irreversibility

Continuing with an account of the influence of geophysical thinking on irreversibility, the next major step was initiated by Charles Lyell. By 1830 scientific speculation was becoming increasingly free from religious dogma. Thus, scientists were considering estimates like millions of years for the age of the earth. There was also an interest among geologists to become "more scientific" in the manner of physics and chemistry. Lyell's "uniformitarian" approach aspired to explain all present geological features in terms of currently operative physical causes. The old problem about the constancy of the internal heat of the earth continued to be a focus of debate. In 1808 it had stimulated the work of Fourier, and by the 1840's and 50's the debate had a different complexion and would culminate in William Thomson's 1852 statement, "On a Universal Tendency in Nature to the Dissipation of Mechanical Energy."

The sequence was this: William Hopkins, tutor of Thomson and Maxwell, argued in the 1840's that existing astronomical data regarding the precession and nutation of the earth-moon-sun system required that the crust of the earth be at least one-fifth of its radius. As a result of his claims, the significance of the "central fires" was diminished and geologists turned their attention to relations between geographical and climatic changes. Hopkins, however, still believed that the cooling of the earth is important if one considers a long enough time scale; there is an overall irreversibility and no evidence of uniformity in the short run can refute this.

Enter Thomson. At an early age he was the recognized British expert on Fourier's theory of heat conduction and a brilliant theoretical physicist. After graduating from Cambridge he spent some time in Regnault's Paris laboratory studying the properties of steam. His inaugural dissertation at Glasgow applied Fourier's equation to the problem of heat distribution inside the earth. Thomson's 1852 statement on irreversibility clearly anticipates the formulation of the conservation of energy and Clausius' rendition of the second law:

1. There is at present in the material world a universal tendency to the dissipation of mechanical energy.

2. Any restoration of mechanical energy, without more than an equivalent of dissipation, is impossible in inanimate material processes, and is probably never affected by means of organized matter, either endowed with vegetable life, or subjected to the will of an animated creature.

2.3 Dissipative Processes and Irreversible Processes Not Yet Distinguished

The period between Thomson's (later Lord Kelvin) 1852 statement and Clausius' 1865 paper containing two fundamental theorems of the mechanical theory of heat (the energy of the universe is constant, and the entropy of the universe tends to a maximum) is a complex and confusing one. For our purposes it will suffice to note that implications for irreversibility in Carnot's theory were not readily grasped by Clausius or Thomson, that "dissipation" and "wasted" were loosely used and vaguely understood terms not clearly distinguished from irreversibility. Thus, the mixing of two kinds of molecules at constant temperature, a fundamental kind of irreversible process, was not even mentioned by Thomson when he listed distinct dissipative processes. Further complicating the clear articulation of a principle of irreversibility was the lack of a principle of energy conservation and a grasp of the convertibility of heat and work. It seems that early discussions were dominated by Carnot's studies, and conceptual advances could come only with quantitative formulations based on more general models of gas behavior.

Clausius studied (1854) various cyclic processes for a gas and distinguished reversible ones that could be run backwards from nonreversible ones. He defined the "equivalence value of a transformation of work into the quantity of heat Q, of the temperature t," as $Qf(t)$ where f is an unknown function of t. If N is the total value of all transformations in a cycle, then $N = \int f dQ$. He was able to show that $N = 0$ for any reversible cyclic process by appealing to an irreversibility principle of the following form: if N were negative, then heat would be passing from a colder to a warmer body without compensation and if N were positive and the cycle were reversible, it could be run backwards again yielding a negative N in violation of the principle that heat can never pass from a colder to a warmer body without some other connected, concurrent change.

By 1865, partly in response to criticisms by Hirn and Rankine, Clausius sharpened his ideas and introduced the term "entropy" in place of the awkward phrase "equivalence-value of a transformation ....." This new concept allowed a simple description of the directionality of cosmic processes, "the entropy of the universe tends to a maximum."

This somewhat grandiose statement had the appeal of simplicity but when we consider the model it was based upon and the looseness of the reasoning, it is not surprising that the interpretation of this statement will become problematic. Clausius offered no satisfactory theory of irreversibility nor a clear distinction between dissipative processes and the more general class of irreversible processes. Indeed, the illustrations he gave for nonreversible cyclic processes where $N > 0$ all involve the production of heat (Clausius, 1867, pp.134-135). With the entry of kinetic theory and statistical methods, things can only get more difficult. Each additional layer of abstraction obscures the locus of the problem of irreversibility; there are more and more ways to go wrong, to be misled by mathematical formalism. Eventually, problems in the foundation of the subject make themselves felt at all levels of abstraction and make possible the sharpest formulation of these problems as paradoxes.

2.4 Statistical Notions Enter Kinetic Theory

Naive statistical notions had already been brought into kinetic theory by 1865 but their development had not reached the level required for the Boltzmann reduction of the Second Law to mechanics. Before we get into the Boltzmann theory, let us digress to briefly trace how statistical notions entered kinetic theory. Of special interest is the ambiguity surrounding the idea of "randomness."

Kinetic theorists before 1856 who identified heat with molecular motion had varying models of just what this molecular motion was like. Some talked about molecules moving with "perfect freedom" and at the same time seem to have held the view that atoms vibrate in a periodic fashion. Others, e.g., Kronig, thought that gas molecules move at constant speed until they strike the irregular (at the molecular level) wall of the container; the resulting complete irregularity of motion was thought to lead to complete regularity of behavior because of the laws of probability. Clausius too used rudimentary statistical methods but it was not until Maxwell that kinetic theory made full use of concurrent developments in probability theory.

Maxwell, apparently influenced by Herschel's review of Quetelet's books on statistics, applied the law of errors to the problem of molecular velocities. Using a law of normal errors derived by Herschel for the case of a ball dropped so as to hit a target, Maxwell (1860) assumed that the x-component of the velocity of a molecule was independent of its y-component. A later treatment (1867) made explicit use of molecular collisions and assumed that the number of collisions of molecules having components v_x and v_y of a certain velocity vector is proportional to the product of the numbers of molecules having those velocities separately. Ultimately, this turned out to be a crucial assumption because it made possible the description of an irreversible time evolution of the velocity distribution functions (by Boltzmann). Maxwell's assumption is in effect an assumption of randomness or molecular chaos and it led eventually to the kind of irreversibility implicated in Boltzmann's H-theorem. Thus, we have the beginning of the connection between "randomness" and "irreversibility."

It is also important to note that in the same year (1867) Maxwell began his correspondence with P.G. Tait, a correspondence in which Maxwell proposed the "demon" who could by judicious sorting of velocities, violate the Second Law. As a consequence, the law was described as having only statistical certainty. In the course of this discussion with Tait, Maxwell fixed upon mixing of different molecules as the fundamental irreversible process. This was to be important later when Boltzmann undertook his investigations of entropy and disorder.

One final observation about Maxwell's thinking: If one follows his remarks on randomness from about 1868 to 1875, there is an aura of ambiguity surrounding them. Sometimes he says that it is only our ignorance of precise data which necessitates statistical methods. Nature is not intrinsically random. Yet other times he seems to talk about individual molecular motion as if it were stochastic, randomness being a property of molecular motion rather than a consequence of our lack of complete dynamical information.

One can see here the roots of later discussions about quantum mechanics where the principle of indeterminacy is distinguished from a principle of uncertainty. The former states that nature is intrinsically "fuzzy;" for the latter the probabilistic character of quantum mechanical descriptions manifests our lack of complete information.

2.5 Boltzmann Tries to Reduce the Second Law to Mechanics

In this section we will give an account of Boltzmann's efforts to reduce the Second Law to mechanics. His work from 1866 to 1878 yielded the unique concept of positive entropy change as a movement to states of increasing probability, methods for handling processes of irreversible diffusion, and the conceptual framework which all subsequent discussions of irreversibility adopted.

Boltzmann's first effort (1860) to reduce the Second Law to the principle of least action resulted in a theoretical proof of the Dulong-Petit law for the specific heat of solids. A paper by Clausius (1871) which contained identical equations prompted a reply in which Boltzmann pointed out in a good natured way the priority of his results. Clausius, in a not so cordial response, claimed that Boltzmann had ignored the difficult problem of a variable force function. As an objection to Boltzmann's 1866 paper, this was irrelevant since the force function never entered the discussion. Only when the force function is introduced does a question arise as to whether a change in the force function can be equated with work. Clausius tried to solve this problem in a rather simplistic way and he did introduce the concept of an instantaneous change in the force function for a system of particles. This idea along with successful probability considerations led to the Boltzmann reduction in 1871.

The sequence was this: Boltzmann's attempts at reduction were hampered by the restriction to closed periodic paths. Under this condition, Hamilton's principle leads to an entropy differential of the form $d\ell n(\sum ET)$, where T is the period and E is the kinetic energy of a particle. The sum is over all particles. Boltzmann sought an alternative expression for entropy which would still be a perfect differential but which was not restricted to a mechanics of closed periodic paths. Therefore, he proposed a formulation based on the notion of the fraction of time which a system spends in a certain velocity and position range. The average value for any function is then simply the sum of the values of the function for each element of phase space times the fractional time the system probably occupies that region.

Boltzmann then went on to consider the case where the force function does not change but heat is added to the system. He interpreted the effect of the added heat as causing changes in the probability that the system dwells in any given element of phase space. In the case where the force function changes, Boltzmann argued that any variation of force function should be substracted from the total variation in order to express the work done. But, by adopting Clausius' idea of an instantaneous change of force function (for which there is no displacement), Boltzmann was able to eliminate any change in force function from the expression for work done.

By this technique he was then able to derive a complex expression for entropy which yielded a simple equation for the entropy of a monatomic gas and also the law of Dulong-Petit, which he had derived earlier.

2.6 The "H" Theorem and Loschmidt's Reversibility Paradox

The following year (1872) Boltzmann published his foundational paper "Weitere Studien über das Wärmegleichgewicht unter Gasmolekulen." He assumed there are no correlations between two molecules before a collision, allowing the joint probability to be expressed as a product of single-molecule probabilities. This assumption has come to be known as the "Stosszahlansatz," which is essentially the same as the modern assumption of "molecular chaos." Using that he was able to show that Maxwell's distribution for molecular velocities corresponded to an equilibrium state. The interesting question now is whether a distribution which is not initially Maxwellian will tend toward a Maxwellian distribution.

Boltzmann first focused attention on the kinetic energies E of molecules and thought about collisions in which the kinetic energies of two molecules change:

$$E_1 + E_2 \rightleftarrows E_1' + E_2' \qquad (2.1)$$

Since energy is conserved, the sum of the energies must be the same before and after the collision. In an equilibrium system, where these collisions must be proceeding at the same rate everywhere, the rate of change of the density $f(E_1)$ of molecules with energy E_1 must be a sum over all possible E_2, E_1', and $E_2' = E_1 + E_2 - E_1'$ of the corresponding collisions, some creating E_1 and some destroying E_1:

$$\frac{\partial}{\partial t} f_1 = \int dE_2 \int dE_1' \, (af_1'f_2' - bf_1 f_2)b^{-1}\psi \tag{2.2}$$

where

$$f_i = f(E_i) \qquad f_i' = f(E_i')$$
$$a = (E_1'E_2')^{-1/2} \qquad b = (E_1 E_2)^{-1/2} \tag{2.3}$$

The square roots, which are not familiar, are due to the fact that the integrations are over energies rather than velocities; that difference disappeared in later formulations and will disappear in our more detailed descriptions below. The function ψ is a precursor to the more modern collision cross-section.

Then Boltzmann found a remarkable functional form he called E but we will refer to as H, using his later symbol and avoiding confusion with our use of E for energy. In the present context that function can be identified as

$$H(t) = \int_0^\infty dE \; f(E)\left[\ln \frac{f(E)}{E^{1/2}} - 1\right] \tag{2.4}$$

The time rate of change of H can be derived from the time rate of change of f (2.2), and various clever manipulations can be employed to show that H must always decrease in time. Furthermore, it could be shown that H was related to the negative of entropy. Thus Boltzmann felt justified in saying that the groundwork had been laid in preparing for an analytic proof of the Second Law. It is true that the derivation was restricted to dilute gases of monatomic point-atoms, and only binary collisions were considered, and external forces were excluded, not to mention the stipulation of spatial uniformity. But there was a beginning.

In 1875 Boltzmann returned to the problem of the H-theorem and offered a new derivation. This time he allowed for the influence of external forces and concluded that each direction of molecular velocity is equally probable and that the velocity distribution in each spatial element is the same as it would be for certain density and temperature conditions in the absence of external forces. A consequence of this is that the temperature throughout a vertical column of gas is uniform and it was this claim that led Josef Loschmidt to propose the "reversibility paradox."

Two things should be noticed here. First, Loschmidt was primarily interested in the problem of thermal equilibrium under the influence of gravitational force. His remarks on reversibility were somewhat incidental to his much broader concern that the "heat death of the universe" apparently implied by the second law was not inevitable. He even speaks passionately of the "terroristic nimbus of the second law."

Secondly, the reversibility paradox did not originate with Loschmidt but was thoroughly discussed by Maxwell, Tait, and Thomson as early as 1867. Only fragments survive but Thomson's 1874 paper "On the Kinetic Theory of the Dissipation of Energy" contains an extended analysis of the consequences of reversing molecular motion. He notes that if we start with an unequal distribution of temperature, allow diffusion to occur, then reverse the motion of each molecule we will return to the unequal distribution of temperature. But if the reversed motion is continued the system will instantaneously return to "equalization." Furthermore, starting with a gas at thermal equilibrium, there is no way to choose a state whose velocity if reversed would lead to equilibrium.

Let us return now to Loschmidt since it was in response to his criticism that Boltzmann was led to develop his ideas on the relation between entropy and probability.

As mentioned above, Loschmidt was primarily concerned with the problem of thermal equilibrium especially regarding its implications for the heat death of the universe. In putting forth his arguments he worked with the following model: There is a system of particles at rest in a gravitation field and one particle is raised to a height h. It drops, striking another particle obliquely and eventually all the particles are in motion and have the same average kinetic energy. Loschmidt then asks whether the kinetic energy in every horizontal plane will be the same. He argues that this is not the case and that h is a limit at which the average kinetic energy is zero.

The second phase of the argument is a denial of the concept of stationary states. Loschmidt supposes the sudden reversal of the velocity of all the atoms so that it would appear that we were standing at the beginning of a stationary state. However, after a long time the state would deteriorate until a single atom would absorb the entire kinetic energy of the system, rise to the top of the vessel leaving all the others motionless at their original place at the bottom of the vessel.

Boltzmann interpreted Loschmidt's argument as an objection against the possibility of a purely mechanical proof of the second law. He then proceeded to give an extended discussion of the reversibility paradox and his theory of molecular motions of gases. The significance of initial conditions had been highlighted by Loschmidt, and Boltzmann agreed that it is possible to find special initial conditions which will not lead to succeeding states of higher entropy. However, he notes that there are only very few choices of initial conditions which will lead to the nonuniform distributions required by Loschmidt.

It is this idea which leads to the remark that calculations based on the relative number of different state distributions could be a method for calculating thermal equilibrium. Clearly, Boltzmann had clasped probability theory to his bosom when he said, "No proof that after the passage of a certain time t, the mixture of spheres must of absolute necessity be uniform whatever the initial distribution, can be given. This is already taught by the theory of probability itself since every nonuniform distribution state, even the most improbable, is never absolutely impossible" (Boltzmann, 1909, p.120). In a retrospective paper, he says that the relation of the second law to the theory of probability first appeared when he pointed out in 1871 that no analytical proof would be possible except on the basis upon which the theory of probability is founded.

Unfortunately, he does not go on to say what that basis is and we shall have to explore the development of probability theory to see what Boltzmann might have had in mind. We will deal with this in a later section.

For now we continue the story of Boltzmann's progressive articulation of the second law. Carrying through his probabilistic method for determining thermal equilibrium, he enumerated all possible complexions for a physical system and assuming that the probability of a macrostate is proportioned to the number of corresponding microstates, he was able to derive an expression for entropy which was directly proportioned to this probability. This is, of course, the famous equation $S = k \log W$ which Boltzmann interpreted as follows: from a, for the most part, very improbable initial state, the system will pass through even more likely states and finally reach the most probable state, thermal equilibrium. Irreversibility is here conceptualized as a tendency to move from less to more probable states.

2.7 The Reversibility Paradox Rediscovered

The next act in the drama of irreversibility opens in England around 1887. Researching conditions for the validity of the equipartition theorem, Tait, Stoney and others tried to give quantitative analyses of the approach to equilibrium for a number of special models. This led Stoney to a rediscovery of the reversibility paradox and to claim that the second law is not a true dynamical law. He even suggested that time itself does not exist separate from events.

In 1890 Culverwell, appealing to the reversibility principle (which he interpreted as saying that for every configuration which leads to an equal distribution of energy, there is another which tends to an unequal distribution) claimed no general proof is possible that a collection of particles will tend to the Boltzmann configuration. He went on to speculate that molecular motions might be inherently irreversible while obeying Newtonian laws collectively.

The British Association gave J. Larmor and G.H. Bryan the task of clarifying the second law and its relationship to thermodynamics. There followed a report which did not shed much light on a subject which had already been rather thoroughly thrashed out by Boltzmann and Loschmidt over ten years earlier. Eventually, (1894) Boltzmann attended a meeting of the British Association and contributed an appendix to Bryan's second report. Response to the report focused on specific heats of gases but things took a different turn when Culverwell (1894) writing in Nature asked, "Will someone say exactly what the H-theorem proves?"

As a result of this plea the reversibility paradox was rehabilitated and once again ready to tax the minds and the patience of the scientists. This time the paradox was framed largely in terms of the H-theorem: where does irreversibility enter the proof? Because of this formulation, the discussions of 1890-1895 were more incisive, more sharply focused than those in the 1870's.

Burbury, in work later credited by Boltzmann as the origin of his "molecular disorder" assumption, made clear that Maxwell's assumption that the number of collisions is proportioned to $n_1 n_2$ does not hold for reverse collisions. In other words, one cannot begin by assuming that two molecules have not collided, but must allow them to collide and then reverse the motion and continue to claim they are not "correlated." Burbury used this to show how the H-theorem imports irreversibility by postulating a kind of randomness. He then says that any actual material system is always disturbed and this suffices to distribute its coordinates in such a way as to make H diminish. "So there is a general tendency for H to diminish although it may conceivably increase in particular cases. Just as in matters political, change for the better is possible, but the tendency is for all change to be from bad to worse" (S. Bradbury, 1894, p.78).

Bryan, Larmor and Watson filled the pages of Nature in an attempt to answer Culverwell's question. The latter took all this in and had the last word by reiterating that the H-theorem can only be valid if somewhere or other assumptions about averages, probability or irreversibility are introduced.

The dialogue was brought to a (short-lived) conclusion when Boltzmann replied to Culverwell repeating his view that the second law is based on probability theory and can never be derived from dynamics alone. Boltzmann went on to describe a graph of H vs. t and claimed that while the probability of H decreasing is always greater than of H increasing, it is certain that H must occasionally increase. Further studies by Burbury highlighted the way in which the assumption about uncorrelated motions limited the domain of applicability of the H-theorem to rarefied gases and Bryan added the point that the assumption might always be satisfied by bringing together a hot and a cold body for which we assume no initial statistical correlation.

For his part, Boltzmann embraced the assumption of molecular disorder generously, crediting Burbury and Kirchhoff, but he was not willing to claim that molecular motions were actually of this sort. The assumption was necessary not only for proving the H-theorem but also for calculations of viscosity, heat conductivity, etc., but this did not compel Boltzmann to adopt a realist position about molecular motions. In order to appreciate his position more fully we need to depart from the historical narrative and look at Boltzmann's philosophy of science.

Note, however, that randomness, molecular chaos or disorder, probability, and irreversibility have all emerged as prominent motifs. Yet we do not find careful analyses of these ideas as part of the dialogue. Boltzmann talks as if a clear, fully articulated theory of probability was common intellectual property, and Burbury never pursues a study of randomness and the various ways it can enter physical theory. New directions in research were to command attention and reshape these issues before they could be explored in the context which gave birth to them.

2.8 Boltzmann's Philosophy of Science

In order to get a sense of Boltzmann's philosophy of science we will simply highlight the major shifts in his theoretical and methodological approach to the second law. This should be sufficient to dispel the idea that Boltzmann was a naive realist.

Early in his career he was strongly influenced by his friend and senior colleague, Loschmidt. In the period from about 1865 to the later 1870's the major theme of his work was the development of physics according to an atomistic principle. All apparent continuum phenomena were to be given an atomistic basis and atoms were understood as having simple existence. When the efforts to give an atomistic or "purely mechanical" proof of the second law were unsuccessful, Boltzmann arrived at a view in which probability laws were seen as necessary additions. Thus, he gave up mechanistic reductionism for a modified form of atomism in which the atoms are "real" but can be treated only with statistical methods.

Two major schools of thought about science had important effects on Boltzmann's development. The energeticists, led by Ostwald and Helm, believed that all knowledge could be derived from one principle, the principle of energy. They had no use for what they called "hypotheses" about the structure of matter since they believed these to be inferior to "principles." The second major school, the mathematical phenomenologists, were sharply criticized by Boltzmann for their view that physics should be reduced to algorithms (in this case, differential equations) for connecting number and geometrical concepts. The phenomenologists headed by Kirchhoff specifically warned against forming any pictures or models of physical processes.

Engaging these schools of thought in extended debate helped to move Boltzmann toward his own theory of science. He introduced the idea that the relevant criterion for deciding among various methods and hypothesis is not whether they are true but rather if they hold promise. Will they lead to interesting and useful results? Will they move research along?

Arguing against the phenomenologists, Boltzmann showed that all theories must go beyond the limits of observation and that ultimately all continuum motions must depend on atomistic ones. The most significant shift in his thinking involved the move from atomistic realism to the theory of mental pictures. The essential features of this view are as follows:

All branches of science are to be developed by using mental pictures of a world made up of "atoms." Any principles or laws may be applied to these units and these pictures should be as general as possible. These are free creations of the imagination to be constructed after listing the phenomena to be explained. These units may be units of space, time, electricity, etc. The pictures are to be judged according to the promise they hold for predicting new phenomena and how well they do in accounting for known results.

Boltzmann saw his approach as deductive as opposed to the inductivism of the phenomenologists but his deductions do not begin with axioms which supposedly describe the true essence of things. He saw his ideas as akin to Hertz but rejected the notion that all our "picturing" has to be in accord with the laws of thought. For Boltzmann there are no permanent laws of thought. Such laws must be subject to evolution. Impressed by Darwin, he was willing to entertain the radical idea that the laws of physics and the laws of thought were subject to evolutionary development.

2.9 The Boltzmann-Planck Debate

In this section we want to show how the question of irreversibility was radically transformed by the exchange between Boltzmann and Planck. This debate helped Planck to extablish a new program of research and eventually this shift in his research program led to a new set of problems which captured the interest of the scientific community. The clarification of irreversibility did not offer the excitement and promise of developing a new form of mechanics and consequently was laid aside until Ehrenfest's *post mortem* encyclopedia article of 1911 (Ehrenfest, 1959).

We begin by looking at a sequence of events ranging from May 1894 to May 1895. This period encompasses the meeting of the Munich Academy where Boltzmann took offense at the treatment of gas theory in Planck's edition of Kirchhoff's Lectures on the Theory of Heat and the Oxford meeting of the British Association for the Advancement of Science, which culminated in a series of articles in Nature devoted to the H-theorem. For our purposes, the important aspects of this sequence are the gradual emergence of a definition (indeed, several) of molecular chaos and the beginning of the shift towards a view that probability is intrinsic to nature.

In the discussion below we will first review the ideas of Planck, which were antagonistic toward kinetic theory, and then consider the Boltzmann-Planck debate, and finally try to assess the importance of that debate for the problem of irreversibility.

Prior to the 1894 meeting of the Munich Academy, Planck had made it very clear that he had little regard for kinetic theory. He thought that all of physics should be reduced to the first and second laws of thermodynamics and that all specific hypothesis about the nature of matter should be eliminated. Planck favored a Kirchhoffian mathematical phenomenology which aspired to representing physics in terms of algorithms for connecting numbers and geometrical concepts. Kirchhoff specifically warned against forming any models or pictures of physical processes. Differential equations played a central role in the program of the mathematical phenomenologists.

In 1891 Planck joined Ostwald in defending the macroscopic viewpoint against Boltzmann and even though he was not an energeticist and Planck was decidedly outside the kineticist camp. When, in his edition of Kirchhoff's Lectures, Planck repeated a derivation of the collision integral and assumed that the molecules after a collision may be treated as independent, Boltzmann was quick to note the error. [This was in May 1894 at the Munich Academy meeting.] At the November 1894 meeting of the Academy, Planck replied that he too had noted the error but felt that, as editor, he should be faithful to the manuscript. He offered, however, the following suggestion: Since equilibrium is by definition the state in which the distribution function is stationary and unaffected by collisions, the probability of an inverse collision may be computed in the same fashion as the corresponding direct collision. In other words, the coordinates may be taken to be independent and separable when one is seeking the form of the equilibrium distribution.

In January 1895 Boltzmann made a somewhat weak reply to Planck's suggestion. He simply said that Planck's argument does not show the uniqueness of the Maxwell distribution. One gets the sense that Boltzmann was interested in Planck's idea. If the discussion of the August 1894 Oxford Meeting with their subsequent articulations in Nature the next twelve months are noted, it becomes clear why he must have been keenly interested. For it was these exchanges involving Bryan, Burbury, Culverwell and Boltzmann that engendered the notion of "molecular chaos" and ultimately, the unfolding of the assumptions which underly the H-theorem.

The irony of the situation in late 1894 and early 1895 is remarkable. Boltzmann in his early derivations of the H-theorem used the product of distribution functions to compute both direct and inverse collisions. But as Burbury and the British physicists focused on the assumptions necessary to derive the H-theorem (viz. Burbury's condition A), it must have become painfully clear to Boltzmann that his own criticism of Kirchhoff applied as well to the H-theorem! And even more disconcerting must have been the realization that the suggestion offered by Planck in defense of Kirchhoff was unavailable to Boltzmann because the H-theorem purportedly applies to nonequilibrium situations.

By May 1895 Boltzmann had revised his reply to Planck and published an article in the Annalen which introduced the idea of "molecularly disordered." The frequency of every sort of collision can be calculated from the laws of probability if this condition is met.

Notice the following year in Boltzmann's Lectures on Gas Theory (Boltzmann, 1964). Boltzmann had arrived at a position which defines molecularly disordered in terms of the validity of the equation for collision frequency. In doing so he explicitly excluded those initial distributions which violate the laws of probability; he assumed that the motion is molecularly disordered and remains so throughout all future time.

For our purposes the thing to notice here is that since Boltzmann is dealing with an actual rather than a most probable distribution function, his assumption amounts to a prohibition of configurations which would be allowed mechanically. And this prohibition seems to be absolute in character. If one tries to connect it with Boltzmann's remark that "Kirchhoff, too, places the assumption that the state is molecularly disordered within the definition of the concept of probability," (Brush, 1964), the following picture emerges: Boltzmann seems to take the concept of probability, understood in some sense as a priori, as a condition for doing statistical mechanics. Molecular disorder is the physical embodiment and/or conceptual consequence of this condition.

Clearly, this is an ambivalent position and one which cries out for analysis and we shall deal with probability, randomness, and molecular chaos below. For now, however, let us continue to trace the labyrinthine course of the problem of irreversibility in order to see how what was once a vital issue was neutralized and separated from the mainstream of theoretical work.

The Boltzmann-Planck debate begins with the problem of irreversibility being cast into a new context by Planck. True to his preference for principles and his desire to avoid basing the Second Law on kinetic theory, Planck, in the preface to his Vorlesungen über Thermodynamik, suggested that electromagnetism might be the correct phenomenon on which to found a principle of irreversibility.

Prompted by Lummer and Pringsheim's work on blackbody radiation he wrote a series of papers, "Über irreversible Strahlungsvorgange" (1897-1900) the first of which claimed that irreversible processes cannot be explained by kinetic theory. If point molecules and conservative forces are assumed, the system must be periodic according to the recurrence objection of Zermelo. Planck thought that a resonator which can emit and absorb radiation could introduce irreversibility into the system in spite of the fact that Maxwell's equations indicate that radiation in empty space or reflected from smooth walls is reversible.

Given Boltzmann's encounter with Planck and his previous debate with Zermelo, we might imagine that he would be somewhat piqued by these remarks by Planck. It must have taken tremendous patience to reply in the restrained manner Boltzmann did. He restated his molecular interpretation of irreversibility and said that any variation of this explanation in terms of electromagnetic processes must still make the same kind of assumption. In other words, Boltzmann is saying that you cannot get irreversibility simply by changing from mechanical to electromagnetic processes. Special initial conditions must be invoked by Planck in order to derive an apparent irreversibility.

It is worth noticing here that neither Planck nor Boltzmann distinguished the reversibility or irreversibility of equations, in this case Maxwell's equations. Furthermore, the discussion was putting old wine in new bottles and obscuring, or at least delaying articulation of the crucial concepts of molecular chaos, randomness, probability, and equilibrium. It is not surprising that these notions are perennially problematic and that some of them emerge in new forms to haunt the theory which the Boltzmann-Planck debate engendered.

The next round in the exchange began with Planck maneuvering to avoid Boltzmann's objection. Recall that Boltzmann had essentially invoked a reversibility theorem against Planck by showing that any interaction between resonators and radiation had to be described by reversible equations. Only special initial conditions could give the appearance of irreversibility. Planck now replied to this by describing the particular process he had in mind. The resonator is supposedly infinitesimally small so that the radiation process is asymmetric - the outgoing secondary waves emitted by an excited resonator have an unlimited intensity in the region of the resonator whereas the intensity of the primary exciting wave is finite. Planck claimed that no processes had yet been found in nature where conservative forces give rise to irreversible changes. He thought radiation phenomena might be a source for irreversibility.

Boltzmann was not impressed with Planck's resonator model and observed that even if infinitesimally small resonators were allowed, it would still have to be surrounded by a strong "electric vibrator" in order for scattering to occur. And this would restore the reversibility of the process. Defending kinetic theory against the recurrence objection (Zermelo), Boltzmann said that for systems containing large numbers of molecules the recurrence time becomes enormously long. He then went on to make a kind of _tu quoque_ argument: the equations of electromagnetic theory are based on a purely mechanical model (granted, an artificial one). If electric resonators could give rise to irreversible processes then there would be a contradiction with Poincare's view that such processes cannot be derived from the differential equations of pure mechanics.

Thus, electromagnetic theory in its reliance on mechanics has no advantage over kinetic theory as a ground for irreversibility.

This is clearly a conflict of basic models - the electromagnetic and the mechanical. Boltzmann stressed the common origins of electromagnetic theory and kinetic theory in "pure mechanics." Planck tried to articulate an electromagnetic model which has sufficient internal complexity to allow for irreversibility without having to make ad hoc assumptions. This was in harmony with his research program characterized by the avoidance of hypothesis about matter and a stress on high level principles of nature. It is not difficult to anticipate that Planck will eventually move away from adjusting the underlying mechanical model with its "resonator" and constraints on the conditions of absorption and emission of radiation. He will need a principle of nature to add to the first and second laws of thermodynamics. And Boltzmann will unintentionally help him reach that goal!

Drawing an analogy which brought kinetic theory and the resonator model closer, Boltzmann suggested that a "most probable state" formulation could be developed such that radiation processes could be shown to go, in general, from more ordered to less ordered states. Clearly, Boltzmann is setting out the possibility for a probabilistic, statistical interpretation of radiation. Slowly, Planck is being drawn in the direction of a statistical mechanical formulation of radiation processes. First, he tried to exclude radiation processes which are synchronized with the system hoping, in this way, to retain only those waves which are disordered. These, he thinks, can be demonstrated to be irreversible. However, Boltzmann found a flaw in the alleged proof and Planck, reaching back to his Inaugural Dissertation of 1879, postulated that "natural" radiation is the only kind found in nature. "Natural radiation" as defined by Planck at this state, still depended on details of the mechanical model, that is, on details about the resonator and the manner in which it absorbs and emits radiation. Generally, however, natural radiation is radiation which is not synchronized with the system and consequently does not have periodic gaps.

Assuming that radiation is always "natural" and that there is symmetry between absorbed and emitted waves, Planck showed that a sort of entropy could be defined.

His formula for the entropy of a resonator with vibration frequency ν and energy U is:

$$S = - \left(\frac{U}{aD}\right) \log \left(\frac{U}{eb\nu}\right) \tag{2.5}$$

a formula which is clearly related to Boltzmann's formula for entropy.

How, then, do we assess the Boltzmann-Planck dialogue? In light of Planck's admission that Maxwell's theory must be supplemented by a hypothesis which says that the energy of radiation is distributed in a completely irregular way over the partial vibrations and his recognition that this hypothesis is the equivalent in electromagnetic theory of the molecular chaos assumption in kinetic theory, we must award the match to Boltzmann. Planck did not succeed in deriving irreversibility from electromagnetic theory. He had to deal with the same set of problems as kinetic theory and ultimately had to import irreversibility into his theory.

But if Boltzmann won the battle, Planck was to win the war, or at least shift the battleground so radically that Boltzmann would never again be able to engage him. The allusion is, of course, to Planck's quantum theory of radiation.

Now, we need to assess the status of the problem of irreversibility after the Boltzmann-Planck debate. What has been clarified? What has been obscured? Has the problem been reformulated in any significant way?

Clearly, the Boltzmann-Planck debate highlighted the elusive character of irreversibility. The problems which had dogged Boltzmann could not be easily evaded by simply looking at a different type of natural phenomenon. As Boltzmann showed again and again, Planck needed to make assumptions of exactly the same sort he had made.

But what is remarkable and difficult to understand is the lack of analytical precision in the debate. All the cards had been played; the situation cried out for a few distinctions to be made; a few key questions asked. If Boltzmann or Planck had separated the "irreversibility" of equations from that of processes, if they had tried to categorize <u>kinds</u> of irreversibility, if they had reflected on the special status of equilibrium, the debate might have progressed to a different level.

As it happened Planck's 1906 formulation of the problem in his Lectures made no advance from his view of 1900. What can be said is that Planck in 1906 displays the effects of his encounter with Boltzmann in that he poses the following paradox: considerations (such as probability) which lie outside the field equations are either superfluous or wrong. Since a full electrodynamic treatment would require specifying amplitudes and phases of all Fourier components of the field (an impossible task), and since, in practice, almost all choices of amplitude and phase lead to the same values of what is experimentally determinable, what we need is a special hypothesis which restricts initial and boundary conditions. This hypothesis need only say that certain extraordinary cases do not occur.

It is clear that little advance has been made here and that Planck is no closer to a mechanism for irreversibility than he was in 1900. By 1913 Planck's treatment of irreversibility had changed considerably. In his revised Lectures, "natural radiation" had been renamed "primitive disorder" and now, rather than being a prohibition against physically permissible states, takes the form of a mathematical stipulation of randomness. Planck eventually dropped the electromagnetic "H-theorem" from the Lectures and used the combinatorial method in the style of mathematical phenomenology. That is, once he had made use of the hypothesis of natural radiation to derive a kind of H-theorem and to justify his way of choosing and counting complexions, he tranformed that hypothesis into what appeared to be a purely mathematical, methodological stipulation. Having climbed the ladder of physical hypothesis to the realm of mathematics, Planck tossed the ladder aside and magically remained in the ethereal domain of mathematical physics.

But if the reader will pardon that rhetorical flourish, there is a serious point to be made here. Recall that Planck was an admirer of Kirchhoff and followed him in his preferences for mathematical phenomenology. It seems plausible to view Planck's work in radiation as a progressive movement away from hypotheses about matter or the particulars of a physical situation toward a formulation whose only contingencies would appear in the guise of mathematical principles or laws about probability. And this general style was to find favor with the vast majority of physicists, at least on the continent.

Those in Britain who were not in the phenomenologists' camp struggled briefly with the problem of irreversibility, at least in the form of the molecular chaos assumption. Burbury tried, and failed to discover a difference between Boltzmann's "molekular-ungeordnet" and his own condition A (Burbury, 1899). Since Boltzmann abandoned the conception after 1896 and it was absent from the 1898 Volume II of Gas Theory, it is not surprising that the kineticists were losing interest in molecular chaos.

So, on the one hand we have physicists, primarily on the continent, beginning their adventures in quantum mechanics imbued with the spirit of mathematical phenomenology and much too busy to worry about a "problem of irreversibility" which after all was only a problem for a discredited mechanistic view. On the other hand, we have kineticists, mostly not physicists, concerned with problems of diffusion, viscosity, and general nonequilibrium situations. These people wanted to <u>use</u> the Boltzmann equation and needed to know when and where it was applicable. Their interests were largely practical and thus the problem of irreversibility was not high on their list of problems-to-be-solved.

But the years from 1905-1910 saw a revival of interest in kinetic theory. The quantum theory of electromagnetic radiation, the Einstein-Smoluchowski theory of Brownian movement, and Einstein's work on photoelectricity all sparked interest in the statistical-molecular model. Because the kinetic theory that had been developed for equilibrium phenomena was readily applicable to quantum systems with only minor mathematical adjustments (integrals became summations), one could say that kinetic theory was influential in the development of quantum mechanics.

This, however, ignores the fact that the branch of kinetic theory dealing with transport processes was not being very actively developed. Since this version of kinetic theory depends on a <u>model</u> and on some details about collisions and intermolecular forces, most physicists were content to wait until quantum mechanics gave more information on atomic and molecular structure. Consequently, most work on kinetic theory involving nonequilibrium and transport took the form of mathematical analysis of equations. So, for example, we find Hilbert's work on the Boltzmann equation taking place in the context of Hilbert's general interest in the theory of linear integral equations. He does not mention irreversibility, does not make any detailed calculations of gas properties, nor is he concerned with conceptual issues surrounding mechanical explanation.

In sum, although some aspects of kinetic theory were part of the growth of science from 1905-1910, it was largely the statistical concepts and methods which were seen as relevant and in harmony with the revolutionary, nonmechanistic world view. In this intellectual climate and with the dominance of formalism in physics and mathematics, the problem of irreversibility must have seemed quaint and archaic to scientists of the "modern" era.

2.10 Ehrenfests and the Problem of Irreversibility

For purposes of the present exposition we choose Paul and Titiana Ehrenfest's *Encyclopedia* article of 1911 to mark the embalming of the irreversibility problem. It is interesting to see how this came about.

Felix Klein had originally asked Boltzmann to do the article but he was reluctant because of the complexity of the task. However, when Klein threatened to have Zermelo, a strong critic of Boltzmann, do the piece, he relented. When Boltzmann committed suicide in 1906, Klein turned to Paul Ehrenfest because he was impressed with his ability to present difficult material. Ehrenfest accepted but the article was not published until five years later. If one looks at what Ehrenfest was doing in those years (and earlier) it is not surprising that the *Encyclopedia* article reflects both his loyalty towards Boltzmann and his fascination with the new physics.

As it happened Klein sent drafts to the printer before reviewing the many intricate footnotes which the Ehrenfests later added. Important and *controversial* material appears in these notes but the unsuspecting reader gets a sense that he is reading an elegant clarification of the conceptual foundations of statistical mechanics. The Ehrenfests seem to be showing the logical structure and necessity of Boltzmann's thought; it is as if a careful *post mortem* were being executed. The presentation is genetic and so it seems that the most important aspects come chronologically latest. A great deal is made of the difference, between the "old" (molecular-based) H-theorem and the "new" (ensemble-based) H-theorem.

The Ehrenfests focused their and the reader's attention on irreversibility as essentially tied to the H-theorem. In doing so, they reinforced the idea that irreversibility is a formal property of systems. This eventually led them to consider the role of ergodic theory and thus plays directly into the hands of the mathematical formalists. There is some irony in the fact that the Ehrenfests treated Gibbs' work only in passing primarily because Gibbs did not take irreversibility and the H-theorem to be fundamental to statistical mechanics and thought equilibrium was the crucial concept. In retrospect, we would have to credit Gibbs with the superior insight and cite the Ehrenfests for creating the impression that irreversibility can only be understood by studying things like ergodic theory. When relegated to that domain, the problem of irreversibility appears to be quite remote and of interest only to a few specialists. It is no longer a general structural or systemic problem which calls for clarification of basic concepts.

CHAPTER 3

THE APPLICATIONS

Questions about the nature and origins of irreversibility seemed to have been laid to rest temporarily by the Ehrenfests' encyclopedia article of 1911. The discussion essentially disappeared for twenty years. This may have been due to a general feeling that the Ehrenfests' argument resolved the irreversibility paradoxes. Indeed that opinion is expressed in recent times. For example, Farquhar (1964, p.19) says,

> Moreover, as has been emphasized especially by the Ehrenfests, it is specifically by way of coarse-graining that the reversible and irreversible aspects of the behavior of a macroscopic system are reconciled - irreversibility is associated with that lack of detailed knowledge which may be obtained about the dynamical state of the system.... .

What is more likely, in retrospect, is not that the Ehrenfests' article resolved the irreversibility paradoxes, useful as it was, but that it came in a time when the interests of physicists were focused very strongly in other directions, especially toward quantum theory. There were continuing advances in equilibrium statistical mechanics, which had been intimately involved in the discovery of quantum ideas, and continued to be wedded in a strange way to quantum mechanics, as if that mechanics were tailor-made as the mechanical model for statistical mechanics. But that development was separated from, and free from the paradoxes of irreversibility. Gibbs (1902) had put the equilibrium theory of statistical mechanics into its final form and rendered it free from any paradoxes. It stood there, and still does, as a clue for the resolution of irreversibility paradoxes, but it was not dependent on their resolution.

A few physicists and mathematicians did retain an interest in the Boltzmann gas theory but they were not so much interested in the paradoxes as in the development of the Boltzmann theory toward applications. They came up with results that are still regarded as remarkable and their successes offered considerable support to the Boltzmann theory even though they did not deal explicitly with the paradoxes surrounding the theory. Their success must be regarded as an important clue for the resolution of paradoxes. If we can learn why the Boltzmann theory works so well in applications we may find the origin of irreversibility.

The applications we refer to have to do with equations used in the prediction of gas behavior. The behavior of any gas, or any other fluid, can be described by a set of equations relating the local values of property densities (e.g., mass density, energy density, ...), the densities of fluxes of those properties, and the space and time rates of change of those densities and flux densities. Some of the equations in that master set of equations can be derived without any reference to molecules, and they are reversible, with no hints of the irreversibility that plagues fluid behavior. But they are not sufficient, by themselves, to describe any pattern of fluid behavior. It is the other equations which must be included to narrow the description down to a specific instance of fluid behavior that bring in the irreversible elements. These are often "equations of transport," which relate flux densities to gradients (rates of change along a spatial direction) of property densities.

The supplementary irreversible equations, such as the equations of transport, can be regarded as empirical, but any intuitive justification or quantative derivation depends on the molecular model for matter. They contain parameters such as "transport coefficients" which cannot be predicted without reference to the molecular model.

The applications we refer to are predictions of the transport coefficients for the equations of transport in fluid dynamics as applied to gases. We will consider the details in Chapter 9, but we will try to illustrate the role of those equations here in order to show what was happening during the period (roughly 1910 to 1930) of applications of the Boltzmann theory.

3.1 Transport Rates Determined by Mean Free Paths

Consider, for example, what happens in a gas when the density n(x) varies in the x-direction from high levels on the left (in the x-direction) to low levels on the right. At any point x there will be a net flux of molecules to the right due to the fact that there are more molecules on the left side of x than on the right side, and the fact that anywhere, right or left, there are as many molecules headed to the right as to the left. The net flux at any point x can be regarded as the difference between the flux J_r to the right and the flux J_ℓ to the left,

$$J(x) = J_r(x) - J_\ell(x) \quad (3.1)$$

This difference is known to be proportional to the gradient (derivative of n(x)),

$$J(x) = -D(dn/dx) \quad (3.2)$$

and that makes sense intuitively. Certainly the greater the difference between the densities n(x) to the right and left of x, the greater the flux. This is Fick's law of diffusion, which is needed to complete many descriptions of transport phenomena. The minus sign in (3.2) is needed there so that J to the right can be positive when the gradient of n(x) is negative (higher densities to the left). The proportionality was confirmed by the observations of Fick in 1855.

The burden on the "kinetic molecular theory" in this simple example is to justify the form of Fick's equation and to evaluate (predict from molecular parameters) the diffusion coefficient D. The effort to accomplish this began long before the Ehrenfests' article. In fact, it began at the beginning of kinetic molecular theory, with Maxwell in 1859.

Maxwell realized that the rate of transport of molecules, or, for that matter, the transport of any molecular properties (mass, momentum, energy, ...), depended on the molecular velocities and how they are distributed. We can get at the question of how the velocities are distributed by introducing the "distribution function" f(v); in three dimensions, $f(v_x, v_y, v_z)$. In a sense, f(v) is the number of molecules in a unit volume with velocity v. More precisely, since v varies continuously rather than simply having discrete values $v_1, v_2, \ldots$ we define f as a "density in velocity." In other words, f dv is the number of molecules in our one-dimensional model with velocities in the range of width dv around a precise v. By an argument that is reviewed in a later chapter here (9), Maxwell concluded that the formula for f in a randomized gas (at equilibrium) is

$$f(v) = Ae^{-(m/2kT)v^2} \tag{3.3}$$

where A is a constant. This is the "Gaussian" or "bell-shaped curve" for the distribution of molecular velocities in one dimension. His formula for three dimensions was quite similar, with v^2 interpreted as square of the speed (nondirectional). With (3.3) in hand, it is possible to calculate the average speed of a molecule. In three dimensions this is described by

$$\bar{v} = \frac{\int v f \, d\underline{v}}{\int f \, d\underline{v}} = \sqrt{\frac{8kT}{\pi m}} \tag{3.4}$$

Here the integral (triple integral) is the abbreviation of

$$\int d\underline{v} = \int dv_x \int dv_y \int dv_z \tag{3.5}$$

Knowing the equilibrium distribution of molecular velocities and hence the average speed, one can calculate the average collision frequency experienced by one molecule z, and hence the mean (average) time between collisions τ. To find z we simply imagine a molecule (sphere of diameter σ) moving with the average speed $\bar{v}$ for one second colliding with every molecule in a cylinder of length $\bar{v}$ (one second) and cross-sectional area $\pi\sigma^2$. The volume of that cylinder is $\bar{v}\pi\sigma^2$ and the number of molecules in it is the volume multiplied by the number density n. Thus we have the collision frequency as

$$z = \bar{v}\pi\sigma^2 n \tag{3.6}$$

The mean time between collisions is 1/z, and hence the mean distance traveled between collisions is $(1/z)\bar{v}$, which is known as the "mean free path," which can be formulated as

$$\lambda = 1/(\pi\sigma^2 n) \tag{3.7}$$

This is a very simple formula (it has been corrected in various ways) which has had a lot of influence in kinetic molecular theory.

Once we have the idea of the mean free path and a formula for it, we can return to the intuitive equation (3.1) and derive the empirical equation (3.2). We simply assume that the molecules moving to the right at x came without collisions from $x - \lambda$ and the molecules moving to the left at x came without collisions from $x + \lambda$. The flux of molecules going in any of the six perpendicular directions at any position is n/6 multiplied by $\bar{v}$, which is assumed to have the same value everywhere, so we have (3.1) as

$$J(x) = \frac{1}{6}\bar{v}[n(x-\lambda) - n(x+\lambda)] \tag{3.8}$$

The gradient of n (the derivative in the x-direction) can be expressed as

$$\frac{\partial n}{\partial x} = \frac{n(x+\lambda) - n(x-\lambda)}{2\lambda} \tag{3.9}$$

and thus we have (3.8) as

$$J(x) = -\frac{1}{6}\bar{v}\,2\lambda\,\frac{\partial n}{\partial x} \tag{3.10}$$

Comparing with (3.2) we can pull out the formula for the diffusion coefficient,

$$D = \frac{1}{3}\bar{v}\lambda = \frac{1}{3}\sqrt{\frac{8kT}{\pi m}}\;\frac{1}{\pi\sigma^2 n} \tag{3.11}$$

This kind of estimation of transport coefficients has been used and improved in various ways since it was initiated by Maxwell. Many of the details are available in any modern text on Physical Chemistry (the chapter on transport phenomena in gases). The earlier improvements, especially by Tait (1886) and Jeans (1925,1954, pp.260-267) produced notable progress. And they were starting with a good idea that gives largely correct dependences of transport coefficients on temperature and density. But that theory did not predict accurate values for transport coefficients, bound as it was to the representation of a molecule as a rigid sphere with a certain diameter, so it gave no dependence on molecular structure. It is regarded today as the best "conceptual" approach to transport phenomena, but nevertheless, crude, primitive, "ultrasimplified" (Hirschfelder, 1954, p.9).

3.2 Transport Rates Determined by the Boltzmann Equation

The estimation of transport coefficients which is regarded as the most sophisticated estimation was developed during the period we consider here, especially at the hands of Chapman and Enskog. And this is the application we are especially interested in here.

Chapman and Enskog were independently looking at the Boltzmann equation from a mathematical point of view, wondering how to use it to estimate transport coefficients. The equation was and is, at first sight, a mathematician's horror. It is a "differential equation" (the unknown function f appearing in the equation through its derivatives) and an "integral equation" (the unknown f appearing in integrands), and it is also "nonlinear" (f appearing in the integrand as a product of f functions). It was way out of bounds for mathematical consideration. The theory of the equation had been discussed by one of the most highly respected mathematicians (Hilbert, 1912) without complete success. On the other hand, the equation had been used by Maxwell to predict the viscosity for a particular molecular model. And Boltzmann had used the equation to prove his famous H-theorem. The equation could be used even if not solved or justified.

With all of this in mind, and much more, no doubt, Chapman and Enskog tackled the problem from different points of view and came up with identical results, which were remarkably accurate and lent themselves to variations in the model for the interactions between molecules. Even more remarkable, perhaps, is that Chapman, who continued on to become well known in many areas of kinetic theory, acknowledged the superiority of the Enskog method and adopted that method for his own use.

In 1916, Chapman published accurate formulas for transport coefficients (Chapman, 1916) and in 1917 (Enskog's Uppsala dissertation) Enskog appeared with identical results. Eventually Chapman teamed up with Cowling (Chapman and Cowling, 1939) to present the definitive text on applications of the Boltzmann equation.

And their approach was later generalized to multicomponent mixtures and to better ways to converge their approximation series (Hirschfelder, 1954). But it is what they did (Chapman and Cowling) that dominates the period we consider between 1910 and 1930.

What they did was to abandon the mean-free-path approach developed earlier by Maxwell, Tait, Jeans, and others and to look for more rigorous ways of calculating fluxes directly from the Boltzmann equation. Maxwell himself had long before abandoned the mean-free-path approach in favor of a method based on the Boltzmann equation using a particular model for inter-particle interactions, and what Chapman and Enskog did can be regarded as a generalization of Maxwell's idea.

The "Boltzmann equation," in the absence of external forces, can be presented in one dimension, as

$$\frac{df_1}{dt} = \frac{\partial f_1}{\partial t} + v_1 \frac{\partial f_1}{\partial x} = \int dv_2 \pi\sigma^2 |v_{21}| (f_1' f_2' - f_1 f_2) \quad (3.12)$$

The left side can be regarded as the rate of change of the density of molecules at (x,t) with velocity v_1 and the right side can be regarded as the contribution to that rate of change due to collisions with molecules of other velocities v_2. One could say, with some oversimplification, that this is the second-order chemical rate equation for the process

$$v_1 + v_2 \rightleftarrows v_1' + v_2' \quad (3.13)$$

where molecules with velocities v_1 and v_2 collide to produce molecules of velocities v_1' and v_2' and vice versa. There is a subtlety in the integrations for (3.13) over "reversed collisions" as opposed to "inverse collisions," which has been missed by some authors (Rice, 1959), and we are warned about the details of the mathematical development. But we can say that, roughly speaking, the Boltzmann equation is the chemical kinetics equation for (3.13), regarded as a "chemical reaction," summing over all possible values of v_2. Or we can see (3.12) as an improved form of (2.2), where velocities are used as independent variables in the place of kinetic energies and where f is allowed to vary with x.

What Chapman and Enskog did explicitly was to acknowledge that there is only one unknown function f in (3.12) even though it is evaluated at various velocities v_1, v_2, v_1', v_2', and to expand it in powers of a perturbation parameter, and then to notice that the zero order formula for f is the equilibrium formula, well known, and then to look for the first correction $f_i^{(1)}$, leaving the second and higher-order corrections for later work. The second-order corrections were eventually developed but have not been very important. The first-order results were remarkably good.

Perturbation theories are often obscure because the authors do not show explicitly the symbol for the perturbation parameter and where that symbol is implanted in the starting equation. That obscurity abounds in our subject. Let us say, explicitly, that the parameter is a and it is placed on the left side of (3.12). Then we may expand,

$$f_i = f_i^{(0)} + a f_i^{(1)} + a^2 f_i^{(2)} + \dots \qquad (3.14)$$

The left side of (3.12) is

$$a\frac{df_1}{dt} = a\left[\frac{\partial}{\partial t} f_1^{(0)} + v\frac{\partial}{\partial x} f_1^{(0)} + \dots\right]$$

$$+ a^2\left[\frac{\partial}{\partial t} f_1^{(1)} + v\frac{\partial}{\partial x} f_1^{(1)} + \dots\right] \qquad (3.15)$$

and the right side of (3.12) is

$$J = J^{(0)} + aJ^{(1)} + a^2 J^{(2)} + \dots \qquad (3.16)$$

where

$$J^{(0)} = \int dv_2 \pi\sigma^2 |v_{21}| \left(f_1^{(0)\prime} f_2^{(0)\prime} - f_1^{(0)} f_2^{(0)}\right) \qquad (3.17)$$

$$J^{(1)} = a\int dv_2 \pi\sigma^2 |v_{21}| \left(f_1^{(0)\prime} f_2^{(1)\prime} + f_1^{(1)\prime} f_2^{(0)\prime}\right.$$

$$\left. - f_1^{(0)} f_2^{(1)} - f_1^{(1)} f_2^{(0)}\right) \qquad (3.18)$$

$$J^{(2)} = \dots. \qquad (3.19)$$

Knowing that f must be a function of a, even though we eventually set $a=1$, we can equate coefficients of a^n for every n on the two sides of (3.12). Thus with $n=0$ we have the "zero-order equation" as

$$0 = \int dv_2 \pi\sigma^2 |v_{21}| \left[f_1^{(0)\prime} f_2^{(0)\prime} - f_1^{(0)} f_2^{(0)} \right] \tag{3.20}$$

and the first-order equation as

$$\frac{\partial}{\partial t} f_1^{(0)} + v_1 \frac{\partial}{\partial x} f_1^{(0)} = \int dv_2 \pi\sigma^2 |v_{21}| G \tag{3.21}$$

where

$$G = f_1^{(1)\prime} f_2^{(0)\prime} + f_1^{(0)\prime} f_2^{(1)\prime} - f_1^{(1)} f_2^{(0)} - f_1^{(0)} f_2^{(1)} \tag{3.22}$$

What makes this perturbation scheme attractive and tractable is that the solution of the zero-order equation emphasizes equilibrium. According to Boltzmann's H-theorem it has to be the equilibrium solution of the Boltzmann equation

$$f^{(0)} = n\left(\frac{m}{2\pi kT}\right)^{1/2} e^{-\frac{m}{2kT} v^2} \tag{3.23}$$

Knowing that, and putting that into the first-order equation (3.21), we see Eq.(3.21) as an integral equation which is linear in the unknown $f^{(1)}$ which can be solved by traditional means to find approximate values of $f^{(1)}$. Furthermore, when we find those formulas for $f^{(1)}$ we know they will have to be linear in the left side of (2.21), which is to say, linear in the derivatives with respect to time and space (x) of $f^{(0)}$, and that means linear in the derivatives of n, T, and v_0, which are the only ways $f^{(0)}$ has to depend on x and t. So we know that $f^{(1)}$ is linear in the gradients. The dependence on spatial gradients only can be imposed by using macroscopic equations of change, in zero order, to replace time derivatives by space derivatives. This is the procedure adopted in the Chapman-Enskog method.

By means of such a perturbation theory, we can formulate the fluxes of molecular property ϕ as

$$J_i = \int dv\, \phi_i v \left(f^{(0)} + a f^{(1)}\right) \tag{3.24}$$

and then recognize the zero-order values as zero, and isolate the first-order contributions, which are linear combinations of spatial derivatives (gradients) of properties, as our best estimates of the transport fluxes, and look at those to pick out our best estimates for the transport coefficients. This very complicated scheme has been carried through with all the algebra and arithmetic (see Chapman and Cowling, 1939) with great success.

At this point we must admit that we have rushed through the applications of the Boltzmann theory (Maxwell-Boltzmann theory) at a rate far exceeding the rate in graduate courses in that subject, but hopefully we have brought out a common denominator that underlies the mean-free-path approximations and the Boltzmann-equation approximations. They both start with equilibrium. In the mean-free-path approach, equilibrium was evident in the use of equilibrium average speeds, average collision frequencies, mean free paths. In the Chapman-Enskog approach, equilibrium was more subtle, but it did appear as the zero-order solution which was needed to find the first-order solution. Equilibrium is very important in both approaches.

The reference to equilibrium in the perturbation theory approach of Chapman and Enskog deserves an additional note. In any perturbation theory, when you treat the time dependence as a perturbation, you end up with a theory that is "strange" (called "singular") in a number of ways. The main puzzle is that it does not give time predictions from initial data. Those data are lost. One could argue that the method does not "solve" the equation in the sense of predicting the evolution of the unknown f from arbitrary initial data but rather reaches out and finds equilibrium as an initial condition and then "uses" the equation to estimate the rate of a forced withdrawal from equilibrium during the early stages of that withdrawal. This brings the purpose of the method around to a full agreement with the purpose of the mean-free-path approach, and, as we will argue in later chapters, the purpose of other methods used in nonequilibrium statistical mechanics. The traditional interpretation of the clues provided by the H-theorem and the success of the Chapman-Enskog applications is based on the assumption that these describe the APPROACH to a FUTURE equilibrium. We assume here that they describe the WITHDRAWAL from a PAST equilibrium. Evidently, we argue, a good description of the early withdrawal, when that has been forced by a change in external conditions, is good enough, at the macroscopic level, to get us to the next equilibrium. And focusing on transitions from equilibrium to equilibrium is the proper way to deal with fluid phenomena. A different approach is required for the description of the evolutions of small aggregates of molecules with rare changes in external conditions (reversible) and even large aggregates with sufficiently frequent changes in external conditions (biological), but for inanimate fluids as encountered in nature, we must focus on equilibria and forced withdrawals.

If we have that focus in mind when we review the various approaches which have been developed toward the description of irreversible processes, we find considerable consistency among the various approaches. First, they all refer to some kind of equilibrium. There has never been any successful attempt to predict rates of fluid processes away from equilibrium without referring to equilibrium in one way or another. The reference may be to an equilibrium average molecular speed to be used to estimate collision frequencies, or to a distribution of speeds that is random enough to justify stochastic assumptions, such as the mass-action law or the assumption of molecular chaos. Or it may be a more explicit reference of the type used by Eyring in his theories about rates for fluid processes. Whatever the reference, equilibrium is always there.

Secondly, the equilibrium referred to in predictions of rates is a past or present equilibrium, not a future equilibrium. It is true that the approach to a future equilibrium is often referred to in general formalisms, and there is a strong appeal in that, and even a little bit of merit, as we will see, but the prediction of nonequilibrium rates for fluids comes strictly from a recent equilibrium.

CHAPTER 4

RETURN TO THE PARADOXES

4.1 The Loss of Information

In 1930 the discussion about the irreversibility paradoxes was resumed momentarily by L. Szilard (1929) and G. Lewis (1930). Szilard revived Maxwell's demon, the "sorting" demon described by Maxwell (1871) as a "door" that could open or shut to selectively allow only the fast molecules to pass from a gas on the left side to a gas on the right side. The gas on the right would keep getting hotter and the gas on the left, colder, in violation of the second law of thermodynamics. Szilard argued that the activity of the demon was such that if it were considered in sufficient detail, its entropy would increase by more than the decrease in the gases and there would be no violation of the second law. According to him and Lewis, irreversibility is not the kind of thing that can be defeated by hypothetical devices but rather it is something that emerges forcefully when we pass from the two-way time of physics to the one-way time of psychology, due to nothing more, perhaps, than the "loss of information" that accompanies the passage. That popularization of the "loss of information" idea, valid as it is, and yet finally inadequate in explanations to physics, has led to important cliches in discussions about irreversibility, and it did lead to the useful Information Theory (1949) developed by Weiner and Shannon. Information theory involves the lifting of entropy out of the thermodynamic context and applying the idea to other contexts, with notable success.

4.2 Microscopic Reversibility

An entirely separate development in 1931 arose from Onsager's discovery of the Reciprocal Relations (Onsager, 1931). To describe this development we must focus attention on the flux equations mentioned in Chapter 3, i.e., the "equations of transport." It was known that the flux of any molecular property i, namely J_i, is often proportional to a "force" X_i, which, in turn, is proportional to the gradient of the density of property i,

$$J_i = L_i X_i \tag{4.1}$$

where L_i is the "transport coefficient" for the process. It has been suggested by Enskog and Chapman, and confirmed by experiment, that the flux J_i has contributions coming not only from the force X_i, previously assumed to be the only cause for a flux J_i, but also from contributions (much smaller, in general) from other forces X_j. For example, diffusion occurs not only in the presence of a concentration gradient, but also in the presence of a temperature gradient ("thermal diffusion"). And so the flux law (4.1) must be replaced by the more general "linear law:"

$$J_i = \sum_j L_{ij} X_j \tag{4.2}$$

Thus the transport coefficients form a matrix L_{ij}. What Onsager proved was that this matrix is symmetric,

$$L_{ij} = L_{ji} \tag{4.3}$$

provided the fluxes and forces are defined in an appropriate way. Finding the appropriate definitions of forces and fluxes and invoking (4.3) constitutes the basic part of the theory known today as "irreversible thermodynamics," which was spread around by the wonderful little book by Prigogine (Prigogine, 1955).

What matters to us here is that Onsager's derivation was based on an assumption now known as the assumption of "microscopic reversibility." Thus we have the irony of founding a theory about macroscopic irreversible processes on the reversibility of molecular processes. Evidently the reversibility of molecular behavior is not inherited at the macroscopic level, but it does give rise to a kind of symmetry in the way transport processes are coupled to each other.

4.3 The Role of Recent Equilibrium

Another separate development (1935) was Eyring's first presentation of his theory of absolute reaction rates. Interest was focused on the "absoluteness" of the predicted rates, which included even calculations of "activation energies," previously left to experiment. The activation energies could be found from "potential energy surfaces," and there was some success in that effort. But that part of the theory turned out to be disappointing because of the difficulty of finding potential energy surfaces for interactions between complex molecules. (There is a revival of interest in that part of the theory at present because of advanced computing possibilities, but that is not the part of the theory that is relevant here.) There were other parts of the theory that had continuing success from the beginning (Glasstone, 1941) and we see in those a major clue for identifying the origin of irreversibility.

Briefly stated here, and considered in more detail in later chapters, the clue from Eyring was this: To predict a rate you must begin with the idea of a system initially at some kind of equilibrium. Then and only then can the method of statistical mechanics (the method of equilibrium) be applied. With that in mind, you can then change a constraint of some kind so that the system is no longer at equilibrium, and it leaves equilibrium. The rate of the leaving can be predicted from knowledge about the system due to its previous equilibrium.

Surprisingly, the intriguing clues offered by Onsager and Eyring about the origin of irreversibility did not stimulate interest then or since in a return to the discussion of the early paradoxes. It was not until the 1940s that further interest was expressed, and that came from another direction.

4.4 Molecular Chaos and the BBGKY Theory

The direction we refer to is the development of what is now known as the BBGKY theory, B for Bogoliubov (1946), B and G for Born and H. Green (1949), K for Kirkwood (1946) and Y for Yvon (1935). The immediate goal was to derive and generalize the Boltzmann equation by means of derivations directly from Newtonian particle dynamics with probability entering in only the "purest" way (no interference with the particle dynamics) until it became necessary to throw in an extraneous assumption. In retrospect we could say the goal was to develop a general approach to nonequilibrium statistical mechanics to parallel the general theory of Gibbs (1902) for equilibrium statistical mechanics. In later years the goal came to be recognized as the identification of the extraneous assumption (nonmechanical) that must be added to Newtonian mechanics to insert irreversibility into the scheme.

Out of the work leading to the BBGKY theory came a rich variety of clues and suggestions, many of which were discussed at the International Symposium of 1956 in Brussels on Transport Phenomena in Statistical Mechanics (Prigogine, 1958). Included there was a discussion about a new approach due to M.S. Green (1956) which we think offered the best clue that comes from the BBGKY theory. We will return to these clues below.

In order to begin a brief description of the BBGKY theory we go back to the Boltzmann theory, which is based on a distribution function $f(\underline{r},\underline{v},t)$, where $\underline{r}$ is the position (x,y,z) of a molecule, and $\underline{v} = (v_x,v_y,v_z)$ is the velocity. We could say that f is the probability (density) of finding a molecule at $\underline{r}$ with velocity $\underline{v}$. The rate of change of the population of a group of molecules localized around $(\underline{r},\underline{v})$ as seen by an observer moving with the group is due to collisions that knock molecules out of the group and other collisions that knock molecules into the group. This rate of change is the three-dimensional form of (3.12). Thinking here of rigid-sphere models, we have that form of (3.12) as

$$\frac{\partial}{\partial t} f_1 + \left(\underline{v}_1 \cdot \frac{\partial}{\partial \underline{v}_1}\right) f_1 = J_1 \tag{4.4}$$

$$J_1 = \sigma^2 \int d\underline{v}_2 \int d\underline{k}(\underline{k} \cdot \underline{v}_{21})\left(f_1' f_2' - f_1 f_2\right) \tag{4.5}$$

The subscripts 1 and 2 are used, as before, for convenience, to distinguish the coordinates and velocities of the two particles considered, and f_1 is f evaluated at $(\underline{r}_1,\underline{v}_1)$. The vector $\underline{k}$ is the unit vector from $\underline{r}_1$ to $\underline{r}_2$ and its integral sweeps through angles that allow, for given $\underline{v}_i$, only positive values of $(\underline{k} \cdot \underline{v}_{21})$. The primes replace the $\underline{v}_i$ by $\underline{v}_i'$, which are the velocities coming into a collision that would produce the $\underline{v}_i$ as post-collision velocities. In the Boltzmann theory, the various f functions appearing in the integrand of (4.5) were evaluated at various $\underline{v}_i$ and $\underline{v}_i'$, but always at the same $\underline{r} = \underline{r}_1 = \underline{r}_2$. This was because the slight difference between $\underline{r}_1$ and $\underline{r}_2$ could not be very significant at the macroscopic level. A more precise evaluation was offered by Enskog (1922) and that will help us in our work.

What Enskog did was to see clearly the two terms on the right side of (4.5) as a term counting the collisions that produce particles (spheres) with velocities $\underline{v}_i$ from precollision velocities $\underline{v}_i'$ minus a term counting the collisions between particles of velocities $\underline{v}_i$ that destroyed those velocities. For the first term, since the $\underline{v}_i$ given are post-collision velocities, then $\underline{r}_2$ must be at $\underline{r}_1 + \sigma\underline{k}$, as illustrated in Fig.4.1a. For the second term, where the $\underline{v}_i$ are

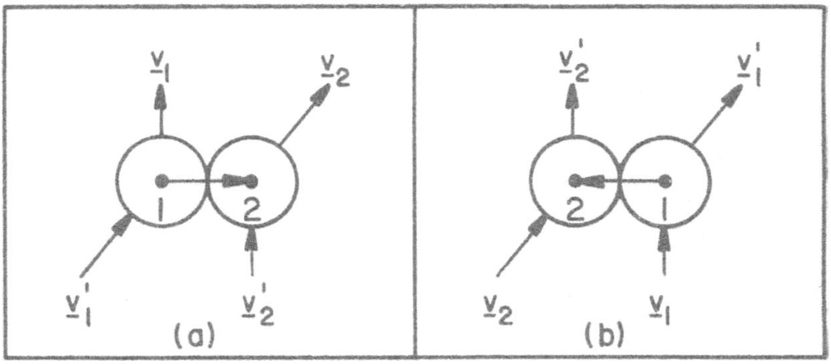

Fig.4.1 Collision (a) and Inverse Collision (b). The vector $\underline{k}$ from the center of particle 1 to the center of particle 2 is opposite in the two cases.

precollision velocities, we must have $\underline{r}_2$ at $\underline{r}_1 - \sigma\underline{k}$ as illustrated in Fig.4.1b. Thus we have the $\underline{r}$-dependence of the f_i in the brackets on the right side of (4.5) as

$$f_1'(\underline{r}_1)\, f_2'(\underline{r}_1 + \sigma\underline{k}) - f_1(\underline{r}_1)\, f_2(\underline{r}_1 - \sigma\underline{k}) \tag{4.6}$$

Since the $\underline{v}_i$ in the second term are precollision velocities and the $\underline{v}_i$ as post-collision velocities in the first term are replaced by precollision velocities $\underline{v}_i'$, it follows that the f function on the right side of the Boltzmann equation is always evaluated at a precollision velocity. This arrow to the past is the origin of irreversibility in the Boltzmann equation. It was not explicit in the original equation and Boltzmann had to argue the point, but it is explicit in the Enskog form described above.

But even if we acknowledge the Enskog arrow to the past (the distinction between inverse collisions and reverse collisions), we still have questions about the reversibility or irreversibility of the Boltzmann equation. If the arrow merely points to a recent equilibrium, as we will propose, then the question can be resolved. But if the Enskog form of the Boltzmann equation is irreversible in time (not invariant under time reversal), then it supports the views of those who think that irreversibility is somewhat mystical next to Newtonian physics. That is the view that brought so much grief to Boltzmann.

There is a definite procedure for testing an equation to determine whether or not it is "invariant under time reversal" (acceptable in the Newtonian scheme of time) which we will consider below. Unfortunately, it is very difficult to carry out the test for the Boltzmann equation and its Enskog improvement. Neither of those were derived directly from Newtonian mechanics, and they both involve an assumption which Boltzmann called the "Stosszahlansatz" about the independence of probabilities of precollision velocities. The derivation from Newtonian mechanics came eventually in a form now known as the BBGKY theory.

The goal of the BBGKY theory(s) was to find a way to derive an exact form for (4.4) and compare that with the Boltzmann equation in order to (1) justify the Boltzmann equation, (2) find corrections, and (3) identify the point at which irreversibility enters the derivation. Thus the first goal was to find the exact precursor to (4.4) with (4.5).

It was known that the exact form of (4.4) could be expressed as an "equation of continuity," preserving the total number of particles for which f is a number density,

$$\frac{\partial}{\partial t} f + \frac{\partial}{\partial \underline{r}} \cdot \underline{v} f + \frac{\partial}{\partial \underline{r}} \cdot \underline{a} f = 0 \qquad (4.7)$$

Here the space is six-dimensional, with coordinates $(\underline{r}, \underline{v})$, and velocity $(\underline{v}, \underline{a})$, where $\underline{a}$ is the acceleration (rate of change of $\underline{v}$), and the second term of (4.7) is the same as the second term of (4.4), since $\underline{v}$ is an independent variable in this context. The trouble with (4.7) as the exact form of what Boltzmann sought is that we do not know the formula for $\underline{a}$ in terms of $(\underline{r}, \underline{v}, t)$. Boltzmann substituted for that intuitive ideas about collisions.

We can, following the BBGKY theory, before it jumps off into new assumptions, start with (4.7) and ask how to formulate $\underline{a}$. We know that $\underline{a}$ is the acceleration of a particle (which we will refer to as particle 1) due to the forces exerted on it by all of the other particles, which we may consider one-at-a-time and then add up their contributions (as vectors). Suppose we have knowledge of the formula for the two-particle distribution function

$$f^{(2)} = f^{(2)}(\underline{r}_1, \underline{v}_1, \underline{r}_2, \underline{v}_2, t) \tag{4.8}$$

which is the probability (density) that particle 1 is at $(\underline{r}_1, \underline{v}_1)$ and particle 2 is at $(\underline{r}_2, \underline{v}_2)$ at time t. And suppose we have selected a molecular model and have the formula for the force $\underline{F}$ on a particle at $\underline{r}_1$ due to a particle at $\underline{r}_2$. The acceleration of particle 1 due to that force is the Newtonian acceleration $\underline{a} = \underline{F}/m$, where m is the mass of particle 2 (all our particles here have the same mass m). Then we can evaluate the overall acceleration on particle 1 due to all possible particles 2 as the sum of all $\underline{F}/m$ over all possible $(\underline{r}_2, \underline{v}_2)$ weighted by the conditional probability that if particle 1 is at $(\underline{r}_1, \underline{v}_1)$ then particle 2 is at $(\underline{r}_2, \underline{v}_2)$, namely $f^{(2)}/f_1$:

$$\underline{a}_1 = \frac{1}{f_1} \int d\underline{r}_2 \int d\underline{v}_2 \; f^{(2)}(\underline{F}/m) \tag{4.9}$$

Thus we have the equation of continuity (4.7) as

$$\frac{\partial}{\partial t} f_1 + \underline{v}_1 \cdot \frac{\partial}{\partial \underline{r}_1} f_1 = - \frac{\partial}{\partial \underline{v}_1} \cdot \int d\underline{r}_2 \int d\underline{v}_2 \; f^{(2)}(\underline{F}/m) \tag{4.10}$$

In the rigid-sphere limit of (4.4) - (4.6), Eq.(4.10) becomes

$$\frac{\partial}{\partial t} f_1 + \underline{v}_1 \cdot \frac{\partial}{\partial \underline{r}_1} f_1 = \int d\underline{v}_2 \int d\underline{k}(\underline{k} \cdot \underline{v}_{21})\left(f^{(2)*} - f^{(2)}\right) \tag{4.11}$$

where

$$f^{(2)*} = f^{(2)}(\underline{r}_1, \underline{v}_1, \underline{r}_1 + \sigma\underline{k}, \underline{v}_2)$$

$$f^{(2)} = f^{(2)}(\underline{r}_1, \underline{v}_1, \underline{r}_1 - \sigma\underline{k}, \underline{v}_2) \qquad (4.12)$$

The comparison between the Boltzmann equation at its best, even though restricted to the rigid-sphere model, (4.4) - (4.6), with the exact equation for rigid spheres (4.10) - (4.12) is where we find the clue that comes from the BBGKY theory. The theory goes on in various ways, some of which will be considered in Chapter 9, to use perturbation methods to find corrections to the Boltzmann equation due to collisions (correlations) among three or more particles. But the basic issue for us is the comparison between (4.4) - (4.6) and (4.10) - (4.12). The Boltzmann equation has the virtue of being an equation for one unknown, solvable in principle, and being "irreversible," but it has the defect of being heuristic and not faithful to Newtonian mechanics. The exact equation has the virtue of being exact and faithful and reversible, but it has the defect of holding two unknowns, f and $f^{(2)}$ and being unsolvable.

To bring these equations together, and to render the exact equation solvable, we have to find a way of expressing $f^{(2)}$ in terms of f. The attempt to do that brings us squarely in confrontation with the problem of irreversibility. We have to bring in an extraneous assumption. What is that assumption and how does it bring in irreversibility to an otherwise reversible context?

Comparing the second terms on the right sides of (4.5), (4.6) and (4.10), (4.12), we see that they become equal if the two-particle distribution function $f^{(2)}$ is a product of one-particle distribution functions

$$f^{(2)} = f_1 f_2 \qquad (4.13)$$

This factorization of probability densities has been cited many times as the probabilistic expression of randomness or lack of correlations needed to derive the Boltzmann equation. It has also been the subject of studies aimed at determining whether the factorization at an initial time can persist in time. Kac (1954) and Prigogine (1962, pp.91,192) have concluded that the factorization can persist for a system which is large enough and dilute enough. We may add that (4.13) is the "most probable" formula for $f^{(2)}$ in terms of f if the only constraints are the normalizations

$$\int d\underline{v}_1 \int d\underline{r}_1 \, f^{(2)} = f(\underline{r}_2,\underline{v}_2,t); \int d\underline{v}_2 \int d\underline{r}_2 \, f^{(2)} = f(\underline{r}_1,\underline{v}_1,t) \tag{4.14}$$

That will be proved in Chapter 9 by the method of Lagrange multipliers.

On the other hand, it is certainly not true that the factorization (4.13) must hold at any time for any gas. We can find infinities of formulas for $f^{(2)}$ that satisfy the constraints (4.14) but do not factor as in (4.13). If we ask for the circumstances under which most probable distributions prevail in natural systems, the familiar answer is: at equilibrium. We are tempted to hurry into the conclusion that the Boltzmann equation becomes thoroughly valid and faithful to the underlying particle mechanics at equilibrium. That is true, no doubt, but it does not help us to understand why the Boltzmann equation works away from equilibrium.

Let us return to Eqs.(4.5), (4.6) and (4.10) - (4.12) and notice, this time, the first term on the right side in each case. Those right-side terms would be identical if we could assume

$$f^{(2)*} = f_1' f_2' \tag{4.15}$$

This factorization differs from the factorization of (4.13) in a very important way. Here $f^{(2)}$ evaluated at post-collision velocities $\underline{v}_1$ and $\underline{v}_2$ is not simply factored into $f(\underline{v}_1)f(\underline{v}_2)$, but rather into $f(\underline{v}_1')f(\underline{v}_2')$, where the $\underline{v}_i'$ are to be derived from the $\underline{v}_i$ as the precollision velocities before a collision that produces the $\underline{v}_i$ as post-collision velocities. The relation between this factorization and the simpler one of (4.13), which was clear enough in the remarkable intuitions of Boltzmann and Enskog without the benefit of explicit reference to $f^{(2)}$, was made explicit in the contribution by Born and Green to the BBGKY theory. The missing link is the fact that $f^{(2)}$ does not have time to change in a rigid-sphere collision. Thus we have

$$f^{(2)}(\underline{v}_1,\underline{v}_2) = f^{(2)}(\underline{v}_1',\underline{v}_2') \tag{4.16}$$

In other words, the probability (density) of having two particles with post-collision velocities $\underline{v}_1$ and $\underline{v}_2$, as in Fig.4.1a, still touching, having just collided, must be the same as the probability of having the same two molecules at the same positions just before the collision. Actually, the assumption of Born and Green was more general, and allowed for "soft" collisions (action at a distance) which take more than an instant of time, and they treated (4.16) as an equation which is not exact but which must be corrected by adding terms to account for the interference of a third particle during the collision period.

If we combine (4.15) with (4.16) we find (4.13) evaluated at the velocities $\underline{v}_i'$, which are precollision velocities. We recall that (4.13) in its original context was evaluated at precollision velocities. So we see that we do not need the factorization (4.13) in general as the extraneous assumption to be introduced to reach the Boltzmann equation from Newtonian mechanics. All we need is that factorization at precollision velocities, never at post-collision velocities. This distinction introduces the lack of symmetry between the future and the past. Somehow it points to the origin of irreversibility.

The assumption described above, namely, that $f^{(2)}$ factors into a product of one-particle distribution functions when (and only when) evaluated at precollision velocities, is now known as the "molecular chaos assumption." It is generalized in the BBGKY theory to the statement that the m-particle distribution function $f^{(m)}$ factors into $f^{(p)}f^{(q)}$, $m = p + q$, when evaluated at positions and velocities such that a trajectory starting there and going backwards in time never reaches a point where particles in one group of p particles interact with particles in the other group (q particles).

Strictly speaking, it may be necessary to include another factor k in the factorization (4.13),

$$f^{(2)} = k\, f_1 f_2 \tag{4.17}$$

This was suggested by Enskog (1922) who used the symbol Y and interpreted Y as a correction that holds at equilibrium. The factor k cannot depend on velocities, for that would negate the whole idea of the factorization, but it can depend on positions. Enskog was thinking of equilibrium statistical mechanics, where (4.17) is exact and does not depend on any extraneous assumptions. It is now known that Enskog's Y function performs very well (Hirschfelder, 1954) and anticipates in a most remarkable way the correction of (4.13) that comes from the BBGKY theory, as we will see in Chapter 9. And there is further confirmation from computer-simulation experiments for a hard-sphere model developed by Alder and Wainwright (Prigogine, 1958, pp.97-131).

The dominant contribution to the BBGKY theory came from Bogoliubov (1946). He offered a perturbation theory that permits the evaluation of any $f^{(s)}$ in the hierarchy of distribution functions from a knowledge of higher $f^{(m)}$ and he offered a set of extraneous assumptions that would "break the chain" and allow $f^{(s)}$ to be determined from lower $f^{(k)}$. Thus every $f^{(s)}$ could be determined from the lowest $f = f^{(1)}$. This provided an expansion for $f^{(2)}$ in (4.10) or (4.11) in a series involving $f^{(1)}$ that included the Boltzmann equations (4.4), (4.5) as a first-order equation. And there was a prescription for the higher-order corrections. The details of the prescription were worked out by Choh and Uhlenbeck (1958), and by Hollinger and Curtiss (1960,1962). The trouble, from our point of view, in retrospect, with the Bogoliubov theory was the complexity and strangeness of the extraneous assumptions. They offered no problems for practical developments, but they presented a challenge to our search for simple origins of irreversibility. Fortunately it has been shown (Hollinger, 1960,1962) that those assumptions which explicitly go against particle dynamics cancel out and leave the assumption of molecular chaos in its purest form as the only extraneous assumption needed for the Bogoliubov scheme.

And so we have the assumption of molecular chaos in the context of the Bogoliubov scheme as our clue for the origin of irreversibility that comes from the BBGKY theory. In that context, the assumption is applied at a past time. We also know that the assumption holds exactly at equilibrium for several particles irrespective of their velocities, which are emphasized so much in other contexts. We are led here to notice Green's "product condition" (Green, 1956). But more we are led to notice the recent equilibrium, which caused all this confusion. Only at recent equilibrium can we resolve all of these puzzles. As we will see, the assumption of molecular chaos and Green's "product condition" are assumptions about recent equilibrium.

4.5 Later Developments

The revival of interest in the problem of irreversibility (reconciling irreversible macroscopic phenomena with molecular reversibility) after the developments described above was substantial. It was influenced by those developments, but many authors searched back to the early beginnings as if there had been no consensus on this intriguing problem (Note 2). They were correct, in our opinion. We still do not know how to explain the simplest examples of irreversibility, as for example, the reason why heat flows from where it is hot to where it is cold, and never the other way. We do know, from kinetic theory, that reversing all the molecular velocities in a stagnant fluid would not change the temperature, which depends on the local average of v^2, and would not change the gradient of temperature, but it would reverse the direction of heat conduction, which depends on the average of v^3. Since we cannot see any reason for the first choice of molecular velocities leading to the correct heat conduction, why should the second choice (reversed velocities, reversed direction for heat conduction) be so impossible? Similarly, friction always decelerates and dissipates, never the other way, even though we can imagine the other way if we are thinking about the motions of individual atoms and molecules. Living organisms always get older, never younger. That lack of symmetry is not present in our picture of the lives of individual molecules. What "force" prohibits the reversed processes in molecular aggregates without interferring in the lives of individual molecules?

The issue has been focused in recent times as the mystique of the second law of thermodynamics, with its predictions concerning the behavior of entropy and its implications about time's arrow (Note 3). The second law would be just another empirical law for the behavior of macroscopic matter if not for the molecular model of matter. There would be no questions, no paradoxes, if not for that underlying molecular model. The model itself has been very successful, even as a basis for statistical descriptions of irreversible phenomena, in spite of its intrinsic reversibility, so it is only fair to ask the model to remove our confusion about the origin of irreversibility in the collective behavior of molecular aggregates.

The issue has also been treated as an exaggeration. According to Grad (1967) there is no paradox. The origin of irreversibility is simply a weakening with time of the dependence on initial conditions, and the barrier between the reversible molecular dynamics and the macroscopic fluid dynamics is not really a barrier to their mutual correctness and compatibility. This can be correlated with Boltzmann's concept of nature "pulling" the system into more and more probable states through a mesh that maintains the Newtonian dynamics. And with Prigogine's concept of irreversibility as no more than the manifestation of molecular randomness on the macroscopic scale (Prigogine, 1980, p.176). According to de Beauregard (1970), irreversibility can be understood by merely recognizing it as "fact-like" rather than "law-like." This is consistent with the suggestions by Lewis (1930), Watanabe (1951), and Mehlberg (1961), although Mehlberg pointed to a difference between local time anisotropy and universal time anisotropy. But de Beauregard does insist on a reconciliation between subjective and objective views of irreversibility. The "loss of information" idea is not enough.

The reactions to the BBGKY theory have been diverse. The method of "time-smoothing" used by Kirkwood and his followers (1946) has been abandoned. Kirkwood argued that the factorization (4.13) could be valid for distribution functions only if they were time-averaged over an interval at least as large as a typical time between collisions. And his procedure did lead from (4.10) to a form of the Boltzmann equation. But it could never have shown the detail of the Enskog equation (4.10) - (4.12) for rigid spheres, and time-smoothing killed off any chance of corrections to the equation.

The extended perturbation theory due to Bogoliubov, explained by M. Green (1956), and pursued for details by Choh and Uhlenbeck (1958), followed by Hollinger and Curtiss (1960,62), has turned out to be the accepted expression of the BBGKY theory. It does offer the Boltzmann equation, even the Enskog equation, as a two-body approximation and goes on to identify corrections due to three and higher-body interactions and correlations, all based on "extraneous" (not Newtonian, not from probability theory) assumptions which can be identified, evaluated, and criticized. Those assumptions, as introduced by Bogoliubov (1946), form a strange mixture of assumptions, but the ones that are deliberately anti-Newtonian cancel out and leave one basic assumption which can be described as the purest form of the molecular chaos assumption.

The assumptions of Bogoliubov have been described as tantamount to the assumption of molecular chaos (Hollinger, 1960), and they have been described as a boundary condition (Cohen, 1962, p.128), but, in retrospect, they could be described better as a consequence of M. Green's product condition (Green, 1956). In our opinion, developed in this book, the best description requires the context of the Bogoliubov perturbation theory with Green's product condition and the acknowledgement of recent equilibrium in the history of any observable fluid. Then there are no "extraneous" assumptions or arbitrary boundary conditions in the BBGKY theory and it can stand next to the Gibb's theory of equilibrium statistical mechanics, and stand there not only as a near equal, but also as an extension which borrows substantially from the equilibrium theory.

The modern view of the Bogoliubov theory (Brush, 1972) is that while it does offer a recipe for finding a series of successive corrections to the Boltzmann equation, nevertheless, that series does not converge. The times involved to go backwards along trajectories for many-particle aggregates to reach times when the particles are not correlated by previous collisions become too large too fast. Our answer to that is that the backing-up along trajectories can stop at the time of the recent equilibrium and then equilibrium statistical mechanics can be used to finish the derivation. After all, equilibrium must come into the derivation somewhere, for only at equilibrium can we justify the use of statistical mechanics, the randomness needed for molecular chaos, for the search for "most probable" distributions. There are hints of inheritance from that recent equilibrium in the "virial expansion" nature of the early results now available for the Bogoliubov corrections, not to mention the hints of confirmation for the Enskog Y function.

Other approaches to the explanation of irreversibility and the prediction of macroscopic evolutions were introduced during the 1950's and 1960's. These came out of discussions of the BBGKY theory. The school of Prigogine (Brout, 1956; Prigogine, 1958 and 1962; Balescu, 1963) and many others (e.g., Cohen, 1962) felt the need to assume that the system must be infinite in some way to have irreversibility. It was well known that the formulas of equilibrium statistical mechanics are "asymptotic" in the sense of holding exactly only for infinite systems. This was an expression of the "law of large numbers" in probability theory, which can be justified by showing that deviations from asymptotic formulas, i.e., "fluctuations," for collections with Avagadro's number of particles, must be so small that we would have to measure to ten significant figures to detect them (Hill, 1960, Chap.2). And so it was natural to look for infinities in the origin of irreversibility as if that were an "asymptotic thing."

There were many other approaches to the "many-body problem" from the viewpoint of quantum mechanics which went beyond the naive assumptions about irreversibility coming from "uncertainty" or any special feature of quantum mechanics as a replacement for the underlying Newtonian molecular model. The Schroedinger equation for many particles had been developed in unpublished works by Pauli and that was revived and refined by Van Hove (1955-1957) and others. Sophisticated methods were developed for approaching the problem of irreversibility (Prigogine, 1956; Cohen, 1962, pp.157-172), but it is interesting to notice that later popular texts (e.g., Fetter and Walecka, 1971) on the quantum theory of many particle systems do not wrestle with the problem of irreversibility. Even the "projection operators" introduced by Zwansig (Brittin, 1961), which have been valuable in other ways, have faded in the literature on irreversibility. Zwansig looked for a common denominator between the classical approach of Prigogine and the quantum approach of Van Hove, and he found it, but there was no way for any of those formalisms to answer the questions about irreversibility raised here. Evidently you cannot "project" irreversibility as something that comes up when small terms fade out in perturbation theories. Some authors went so far as to see irreversibility coming out of "resolvent operator formalism," distinguishing between discrete eigenvalues and those that form a continuum. We have no doubt that what they were doing was valuable in many ways, but we also have to notice that modern authors in those areas do not even have "irreversibility" in their indices.

There was some attention in the 1950's to the possibility of "hidden variables" as the cause for mysterious things such as irreversibility, or "measurement problems" as the cause, as if those issues were not transparent in both quantum and classical mechanics. And other possibilities were mentioned, such as the restrictions imposed by special relativity, by unnoticeable perturbations which would interfere with the Poincaré recurrence (periodicity), etc. And there were many arguments suggesting that irreversibility comes from the coarse-graining techniques in statistical descriptions, and other techniques, as if the irreversibility of nature could care less about ways of describing it. In 1970 the problem of irreversibility stood pretty much where it was a century before (and may stand a century ahead) except for a lot of formalisms, with very little consensus.

There is today an indifference to the problem of irreversibility, which could be due to the fact that the problem is not very important. After all, applications of theories have proceeded without the interference of the problem. But it is possible that a resolution of the problem would provide guidelines for the development of nonequilibrium theories beyond their present stages. In any case, it would be nice to understand where irreversibility comes from. The one feature of irreversibility we can be sure of is that it is not a fine detail which could arise from a hidden or subtle cause, or an illusion, but it is a gross and urgent feature of systems which seem to be described very well in terms of the molecular aggregate model. We are compelled to believe that the origin of irreversibility is a strong and recognizable feature contained within that molecular model.

CHAPTER 5

VARIOUS KINDS OF IRREVERSIBILITY

One of the obstacles in the path toward understanding irreversibility is the confusion that arises from failing to distinguish among the various meanings of the terms "reversible" and "irreversible." In fact, it is possible that the paradoxes of the past and the near-mysticism of the present (Prigogine, 1980) may be due primarily to a lack of interest in distinctions among the various kinds of irreversibility at an elementary level. This is not to say that the arguments were weak or the issues irrelevant, but that the resolution of paradoxes and the reconciliation of Newtonian molecular behavior with the irreversibility of fluid behavior may become easier if we do bother to sort out the various kinds of irreversibility.

We have to say, in retrospect, that it is surprising how much heavy reasoning could pile up around the "problem of irreversibility" without first striving for a clear idea about the precise nature of the problem. We must also express some concern, in retrospect, that the results we get from sorting out the various kinds of irreversibility may be an over-reaction to the gain from that sorting. In any case, the sorting is necessary.

In this chapter we will try to sort out the different kinds of irreversibility, some of which are quite independent of each other. The classifications that are explicit here will be carried over to later chapters. Our goal, however, is not so much the sorting as it is the identification of the kind of irreversibility which needs to be explained and to offer that explanation.

5.1 Inertial Irreversibility

Consider first the kind of irreversibility referred to most often in texts on thermodynamics, and in chemistry generally. This is not the kind of irreversibility we will emphasize, but it is important and we have to recognize it to avoid possible confusions.

Entries under "Reversible" in a journal index, or usage of that term in textbooks and monographs on thermodynamics or physical chemistry (the term is "quasistatic" in engineering texts), refer to macroscopic processes that occur while the system remains near equilibrium. The process is called "reversible" because it can be reversed at any stage by a slight change in external influences. In the case of an expanding gas, for example, the gas pressure is equal or nearly equal to the pressure of the fluid outside. In the case of a phase change at the equilibrium temperature and pressure, the process is "reversible" because it can be reversed by the slightest change of the outside pressure or temperature. Heat flow is reversible if the temperature gradient is next to negligible. Electrodes are reversible if the current can be reversed by a slight change in the potential difference (Note 4). The term "reversible" (and, similarly, "irreversible") as used in this sense, is very common and very well explained. (See, for example, Fast 1962, pp.12-14).

As considered here, a process is reversible because it can be actually reversed in a short time by a slight change of external influences. Strictly speaking, again, the reversible process has an infinitesimal rate and can be reversed in an infinitesimal time by an infinitesimal change of external influences (i.e., external forces, or temperatures of reservoirs, or pressures, etc.). Practically speaking, many interesting processes are slow enough and the system remains close enough to equilibrium so that the process can be treated as reversible. The restriction to a slight change of external influences is not the key factor, for it enters only because of the neglect of "inertia" in the thermodynamics of uniform, stationary systems. If we allowed for inertia, we could allow for arbitrary changes of external influences. The process would be reversible only if it could be reversed instantaneously, or nearly so, by any change of external influences, short of an infinite change.

The reference to inertia needs clarification. In the case of fluid flow the inertia can be taken into account explicitly without difficulty, as in the case of particle motion, but in the cases of transport phenomena (conductions), phase changes, chemical reactions, etc., the idea of inertia is an analogy similar to the analogy of the generalized forces of irreversible thermodynamics (e.g., a temperature gradient being a "force" for heat flow). (See for example Prigogine, 1955. For generalized forces, see p.29). On this analogy we can say that a process is reversible if it has no (nearly no) inertia.

A paradox arises if we extend the idea of reversibility identified above to the realm of particle motion, where the "process" under consideration is not a macroscopic phase change or anything like that, but merely the motion of a body or the motions of several bodies, whether they are molecules or basketballs. Ideally, the only reversible body or particle is one that is not moving. Motion equilibrium for macroscopic bodies or for molecules means zero velocity, no inertia. For a single particle this criterion of reversibility is meaningless; it depends on the selection of a frame of reference, but for an aggregate of particles we can say that their "process" is reversible if they are not moving (nearly so) relative to each other. Even then, the classification raises the question of how to classify a beam of particles moving in the same direction with the same speed, a very "cold" beam with no "spread" (temperature) of particle velocities, but possibly moving fast in some reference frame. We could simply say that such a process is reversible in some frame of reference.

The commonly encountered fluids, regarded as aggregates of molecules buzzing around in all directions, have to be classified as systems undergoing vigorously irreversible processes at the molecular level. The individual molecules have inertia. At the same time such aggregates may appear to have no motion, to be undergoing "reversible processes," at the macroscopic level. Thus, we are led to ask how reversibility at the macroscopic level can emerge from irreversibility at the molecular level. But if this is a paradox, it is opposite to the paradox we seek, which compares reversibility at the molecular level with irreversibility at the macroscopic level.

The resolution of the paradox of the emergence of macroscopic reversibility from microscopic irreversibility (i.e., emergence of fluid equilibrium from molecular motion) has been developed extensively in the literature of chemical education (e.g., Carmody 1960). The concept involved is referred to as "dynamic equilibrium" and the mechanism is the cancellation of opposing molecular processes (Hollinger, 1972).

The first kind of irreversibility expresses the idea that a process is reversible only if the system remains unchanging at equilibrium or changes slowly while remaining near enough to equilibrium to justify the approximate use of equilibrium ideas and formulas in the description of the system.

Criterion 1: A process is reversible if it occurs so close to equilibrium that it can be reversed quickly by changing the external influences (no inertia).

5.2 Temporal Irreversibility

There is a second kind of irreversibility, which is also not the one we seek, but it must be noticed for it has been presented in monographs as the kind we seek and there is a great deal of confusion about this kind of irreversibility. According to this second kind of irreversibility all processes are reversible at both the macroscopic and microscopic levels, so there will be no paradox about how one kind of behavior emerges from another.

This kind of reversibility is often called "time-reversibility" or "invariance under time reversal." The reference to "time reversal" means reversing not only time but also the directions of all motion-variables, such as velocities, momenta, fluxes, etc. The basic assumption throughout is that time is intimately connected to motion, and reversing one of these is equivalent to reversing the other. If we reverse both, conceptually, of course, the result is that the evolution proceeds as if we had not reversed either (it is "invariant under time reversal").

This criterion of reversibility is generally illustrated in terms of equations used to describe the processes (Farquhar, 1964). Consider, for example, the Newtonian equation $F = ma$ expressed as

$$m(dv/dt) = F(x) \tag{5.1}$$

where x is the position of a particle of mass m along a line, v is the velocity of the particle in that direction, and F is the force on the particle in that direction. We assume that the force varies with the position but not the velocity v, and it does not vary with time except through its dependence on the position x. (The argument could be generalized.) If we include another equation, which can be regarded as a definition,

$$v = dx/dt \tag{5.2}$$

then we have two equations for the two unknowns x(t) and v(t), so the trajectory (the unknown functions) can be predicted from initial data, x(0) and v(0).

If we look at Eqs.(5.1) and (5.2) and impose a change replacing v by -v and t by -t, assuming that m and F are not affected, we find cancellations of the minus signs which leave the equations unchanged, which is to say that they are "invariant under time reversal." Reversing the direction of time, i.e., replacing t by -t, by itself, would change the equations, and rightly so, for they would have to describe motion opposite to the motion described by the equations in their original form. And reversing the direction of the motion variable v, i.e., replacing v by -v, would have a similar effect. But reversing both the direction of time and the direction of motion leaves the equations unchanged, and leaves the motion unchanged (invariant). The "system" is "reversible" (in time) in the sense of being "invariant under time reversal."

One is tempted to challenge the classification described above by pointing out that Eqs.(5.1) and (5.2) can be combined to form the more familiar expression of Newton's formula,

$$m(d^2x/dt^2) = Fx \tag{5.3}$$

which is not only "invariant under time reversal" but even more "symmetric" in time. It is unchanged by reversing time (replacing t by -t) even if we do not reverse the direction of motion explicitly (replacing v by -v), for the symbol v does not appear in the formula. Such symmetry in time, carried to logical extremes, would go strongly against our expectations. It would imply, for example, that diffusions and dissipations would look the same in both directions of time. But in the test of Eq.(5.3) for invariance under the full-fledged time-reversal, we have to notice that the equation is second order, and hence its solution depends on the initial value of the velocity v. We would have to reverse the sign of that number, and then we would find ourselves back in the ordinary, asymmetric world of invariance under time reversal.

The test described above as applied to Newton's equations for a particle moving in one dimension can be applied to Newton's, Lagrange's, Hamilton's, anybody's equations for any system with the same conclusion; they are all invariant. And similarly for Schroedinger's equation, any relativistic equation, any other equation of motion. The theory of elementary particles has raised new questions, but there are not yet any suggestions there which have to be noticed here. One is tempted to argue that all dynamical schemes for predicting motion are, so far as we know, invariant under time reversal, and so, perhaps, are the equations of fluid dynamics and other equations describing macroscopic phenomena.

Let us apply the test of "invariance under time reversal" to the simplest equations of fluid dynamics. Consider, for example, the "equation of continuity" for a simple fluid moving in one direction, the x-direction,

$$(\partial/\partial t)\,\rho + (\partial/\partial x)\,\rho v = 0 \qquad (5.4)$$

where ρ is the mass density and v is the fluid velocity. Since ρ is unaffected by the (t,v) reversals, the equation is invariant. The same conclusion is easily reached for the other fluid equations that involve time rates of change, i.e., rates of change for the densities of energy, momentum, entropy, enthalpy, etc. But when we consider transport equations and chemical rate equations, we find a different story.

Consider, for example, the Fick equations for diffusion. Take the case where we have one component at concentration c diffusing through a background medium of fixed composition. The flux J of the moving species is related to the gradient of c by

$$J = -D(\partial/\partial x)c \qquad (5.5)$$

and we also have

$$(\partial/\partial t)c = D(\partial^2/\partial x^2)c \qquad (5.6)$$

The flux J is a motion variable and so it changes sign under time reversal. The position x and the concentration c certainly do not change, but the time t does change sign. Under time reversal, Eqs.(5.5) and (5.6) are altered; the minus sign explicit in (5.5) disappears and a new minus sign appears in (5.6). These altered equations are quite different; they describe a world in which diffusion goes the "wrong" way. We are tempted to conclude that the diffusion equations really are "irreversible" in the sense of being not invariant under time reversal and we are tempted to identify that irreversibility with the commonly expressed irreversibility of diffusion,with the idea associated with "decay," "dissipation," "anisotropy in time," etc. (see Liboff, 1969, pp.319-325).

Before we jump to the conclusion just described, let us look again at Eqs.(5.5) and (5.6), and again apply the test of time-reversal, changing t into -t and reversing the signs of motion variables. What happens to the diffusion coefficient D under these operations? This parameter is not a mass or a charge or a universal constant and we cannot assume that D inherits invariance from those more fundamental particle parameters. We have to predict the symmetry of D by statistical arguments involving explicit derivations from particle equations. When that is done it is found that D changes sign under time reversal. Even in the simplest derivations (Jeans, 1954), where D is found to be proportional to the mean time between particle collisions (mean time before the <u>next</u> collision, or the reciprocal collision frequency or mean free path), it is found that D changes sign (Yang and Lees, 1960 and Biolsi, 1974). The same conclusion is easy to reach for other transport coefficients and chemical rate constants. All of the equations of fluid dynamics are invariant under time reversal, as indeed they must be. Irreversibility in the sense of violating invariance under time reversal is logically untenable within dynamical schema. Even the famous Boltzmann equation, upon which the paradoxes about irreversibility were initially erected, and which is still said to be irreversible, is actually invariant under time reversal if we apply the test carefully to the sufficiently explicit form of the equation provided by Bogoliubov (Note 5).

We have described two criteria for reversibility (or irreversibility). Let us restate the first one in a way more comparable to other criteria, and state the second one (time reversal) just discussed, and include the third one we are about to consider:

1. A process is reversible if it can be reversed by changing external influences, i.e., if the system is at or near equilibrium. (Most processes are reversible at the macroscopic level and irreversible at the molecular level.)
2. A process is reversible if it can be reversed (conceptually, in time) by reversing the motion variables. Or we may say that the process is reversible if it can be reversed by reversing time. (All processes are reversible at any level.)

3. A process is reversible if the process and its reverse both occur naturally. (Most processes are irreversible at the macroscopic level even though there is no such irreversibility in the life of a single, isolated molecule.)

5.3 Exclusion Irreversibility

The third criterion is best known in terms of the behavior of entropy. The processes which do occur in nature seem always to have increasing entropy (for the system and its surroundings), and they seem to be able to go only one way. If we believe that all processes are reversible in the sense of criterion 2, then we may add that the excluded processes are those attended by decreasing entropy. There are many examples of this kind of irreversibility: heat flows from where it is hot to where it is cold, and never the other way, even though the other way is as easy to conceptualize in terms of molecular motions. A chemical reaction on one side of equilibrium seems to have to go to that equilibrium, never the other way; the entropy of the world always increases.

If we assume that all processes are reversible in the sense of criterion 2 (reversible in time), then we are forced to notice the exclusion of some time-reversed processes, i.e., irreversibility according to criterion 3, and to look for the reason for the exclusion. The absence of those entropy-destroying processes must be due to external causes, maybe historical constraints that keep systems away from certain configurations. Consider, for example, rainfall. Since the rain usually moves downward and rarely upward, we may say that the process of rain is irreversible in the sense of criterion 3. It is clear in this case that the irreversibility does not arise from a violation of invariance under time reversal of the laws of motion but rather from natural constraints on the way raindrops are formed. This is the kind of irreversibility we are especially interested in, and we want to find its origin in the case of phenomena governed by the second law of thermodynamics.

Before we try to identify the mechanical origin of entropic-irreversibility we make some digressions that will simplify the task. First we emphasize the distinctions among different kinds of irreversibility already introduced by considering "counter arguments" in which the distinct kinds are mixed. Then we will focus attention on the "survivor" worthy of further study, namely, exclusion irreversibility (3), and discuss the relations among various examples, including entropic and statistical irreversibility. We will eventually review the influence of statistical descriptions on the problem of explaining entropic irreversibility in order to understand why the mechanical explanation has been ignored.

5.4 Mixing the Criteria: Thermodynamic Irreversibility

Here we consider the mixing of Criteria 1 and 3. It is especially easy to interchange these two criteria as if they were equivalent or to try to combine them into one complex criterion. The assumption of equivalence or the attempt at synthesis is at least tacit in almost every text on thermodynamics. In some cases reversibility and irreversibility are defined first according to criterion 1 and then later regarded as, or developed into, complex concepts involving (1) and (3) (Fast, 1962; Moore, 1972, pp.34 and 89; Sears, 1956, pp.3,4 and 139-144). Other authors start with (3) expressed as entropic irreversibility and then modify the criterion to include (1) (Landau and Lifshitz, 1958, pp.22-31; Bent, 1965, pp.4,51,57, and 103; Morse, 1964, pp.61,63). The goal in either approach is to end up with a criterion that refers to a process as reversible if the system remains near equilibrium with essentially no entropy production, and refers to a process as irreversible if it is irreversible in the sense of (1) (has inertia) and in the allowed direction in the sense of (3) (is spontaneous) with positive entropy production. The behavior of entropy is supposed to supply the unifying principle.

Strictly speaking it is not correct to identify a process which is irreversible (1) with one that is in the spontaneous (allowed) direction (3), i.e., the direction of increasing entropy. The reverses of spontaneous processes are also irreversible in the sense of (1) even though they may be forbidden under some circumstances (3). Furthermore, a spontaneous process is often initially reversible in the sense of (1) and irreversible in the sense of (3). This is the case initially when a system is forced to leave one equilibrium to approach another; the process can be easily reversed (1) at the initial time, but only if acceleration in the reversed direction is allowed. Inertia and spontaneity are not quite the same thing.

Denbigh describes the confusion about reversibility in thermodynamics and argues that the thermodynamic criterion for reversibility should be defined in thermodynamic terms, with special attention to what is happening to the environment of a system during a process (1971, pp.23-35, 40,41). He distinguishes between reversibility in the sense of (1) and "thermodynamic reversibility," pointing out that the former is "necessary but not sufficient" for the latter.

In our opinion the major cause of confusion about irreversibility in thermodynamics arises from the failure to distinguish between two different ways of reversing a process. Consider, for example, a process in which heat is flowing from a hot body A to a cold body B,

$$A(\text{hot}) \rightarrow B(\text{cold})$$

If we reverse the situation so that A is cold and B is hot, then the spontaneous direction of heat flow is reversed <u>spatially</u>,

$$A(\text{cold}) \leftarrow B(\text{hot})$$

but that is not the same thing as allowing the heat flow to be reversed in the sense of (3),

$$A(\text{hot}) \leftarrow B(\text{cold}).$$

If A and B are near thermal equilibrium, then the flow of heat is spatially reversible, reversible in the sense of (1), but it is still thoroughly irreversible in the sense of (3). It is very easy to forget the distinction between (1) and (3) near equilibrium if we confuse spatial reversibility (1) with the subtler reversibility of (3).

We think the confusion about reversibility in thermodynamics is most easily cleared up by simply recognizing separately criteria 1 and 3 without trying to combine them. We could ignore criterion 2, or include it for emphasis:

> <u>Thermodynamic Reversibility</u>: All processes described by thermodynamics are always irreversible according to criterion 3, always reversible according to criterion 2, and mixed according to criterion 1. At equilibrium they are reversible (1) with no creation or destruction of entropy. Away from equilibrium in the allowed direction (3) they are irreversible (1) and create entropy (3).

5.5 Mixing the Criteria: Paradoxical Irreversibility

It is also easy to bring together, as if equivalent, criteria 2 and 3. Rightly or wrongly the inseparability of these criteria is taken for granted at every opportunity. The assumption is explicit in most studies on topics in statistical mechanics which contrast the "time-reversal invariance" of mechanical equations with the "irreversibility" of other equations closer to the macroscopic description (e.g., the Boltzmann equation, various "master" equations), not to mention the macroscopic equations themselves (Note 6). The assumption shows up implicitly in many ways, as for example, in the use of the term "microscopic reversibility" (e.g., Prigogine, 1955, pp.51,52, and de Groot and Mazur, 1962, pp.92-100), as if reversibility in the sense of (2) were present only at the microscopic level. Because of this unfortunate assumption, it is assumed further that the second law of thermodynamics cannot be explained in a Newtonian framework (Landau and Lifshitz, 1958, pp.22-31), and that assumption is implicitly spread very widely and deeply.

The reason for connecting criteria 2 and 3 is that the separation of <u>allowed</u> and <u>forbidden</u> processes according to (3) seems to agree with the separation of <u>forward</u> and <u>reverse</u> processes (or vice versa) according to (2). Such an overlap of "results" seems to point to a common denominator for the two criteria. And yet, there are differences. Criterion 2 can be applied at any level of description while criterion 3 seems to operate only on the macroscopic domain. Moreover all processes must be "reversible" according to (2) in order that we may be able to recognize processes which are "irreversible" according to (3).

Our main argument against the dissolving of criterion 3 into criterion 2 is the fact that no sure violation of reversibility according to 2 has yet been found while irreversibility according to 3 is our constant companion. It is true, as was mentioned above, that the test of time-reversal has been applied, sometimes explicitly, many times tacitly, with results favoring the irreversible classification (2) of equations at the macroscopic level, but the conclusion is reversed if we apply the test with sufficient care to assign the correct symmetry to every symbol in the equation. All equations which have established roles in fluid dynamics and which have derivations for the constitutive parameters (e.g., diffusion coefficients) from molecular parameters are reversible in the sense of (2) as indeed they must be if the derivations are faithful to the molecular model.

The kind of irreversibility we get by interpreting the observed and statistically predictable phenomena with exclusion irreversibility (3) as being connected with time irreversibility (2) is certainly paradoxical. Zermelo was right; there is a paradox. Time irreversibility for a macroscopic fluid as against time reversibility for molecular motions is a bonafide paradox. This is not like inertial reversibility (1) at the fluid level as against inertial irreversibility (1) at the molecular level, which can be explained as easily as getting small numbers from the subtraction of large numbers. Rather, it is an attack on the very concept of time, which, whatever it is, must be the same, for our sanity, for both molecules and fluids.

Furthermore, the paradox goes even deeper. There is no way to recognize exclusion <u>irreversibility</u> without assuming time <u>reversibility</u>. The fusion of criteria 2 and 3 is not only a paradox but a contradiction. We must agree with Zermelo, backed by Poincaré. The only way Boltzmann could fight the paradox was by distinguishing between criteria 2 and 3, which he may have done, but not explicitly enough.

5.6 Refinements: de Facto and Nomological Irreversibility

Let us now restrict ourselves to the kind of irreversibility which has not been explained, as it has been in the case of criterion 1, or recognized as a general assumption, as in the case of 2, but which remains our source of mystery, namely, 3. This criterion refers to processes whose reverses are somehow excluded. We will always accept irreversibility in the sense of inertia (1) without further consideration. We will maintain our constant assumption of reversibility in the sense of (2) for all processes in all descriptions at all levels, mentioning it only to remind ourselves of its necessary and everpresent role in our reasoning about irreversibility (3).

We referred above to irreversibility (3), which recognizes the exclusion of the reverses of observable processes, as "entropic irreversibility," as if all examples of such exclusions were related to the second law of thermodynamics. But now that we have settled on this kind of irreverisibility for further study let us widen our view to include all possible examples.

First we must distinguish between processes that have reverses which are excluded by natural means and processes that are *a priori* irreversible in the sense that the reverses are conceptually impossible. We will find that this distinction separates the irreversibility of *observed* fluid behavior from the irreversibility of *statistical predictions* of fluid behavior.

There are many examples of *a priori* irreversibility. Consider a particle moving in a vacuum relative to a fixed observer. The particle can come toward the observer or move away from the observer, but eventually it must move away. Or consider a signal originating at a point and moving out into a nonreflecting space. It can go out, but not come in. Or consider a gas of noninteracting molecules in an empty universe. The gas can expand or contract, but eventually it must expand.

Such ideal examples have played a key role in philosophical discussions about the anisotropy of time. Some parts of those discussions are helpful here because they force us to refine our notion of exclusion irreversibility.

A distinction which is often made in the philosophical discussions is that between nomological and *de facto* irreversibility (Note 7). Nomologically irreversible processes are those whose temporal "inverses" (i.e., reverses) are impossible because they are ruled out by a law or combination of laws of nature. An example which has been used to illustrate nomological exclusion is the impossibility of a cigarette reconstituting itself from its ashes.

A process is *de facto* irreversible if the temporal inverse in fact never or hardly ever occurs with increasing time. Particular initial or boundary conditions obtain independently of any law or laws and combine with a relevant law or laws to render the temporal inverse nonexistent. Although no law, by itself, disallows the inverse process, that process hardly ever occurs.

The studied cases of *de facto* irreversibility can be easily cited. In open systems the inverses of certain processes are not realizable by physical systems. Thus the temporal inverse of a spherical light wave expanding eternally from a center through infinite space would require radiant energy coming "from infinity" in a coherent fashion and converging on a point. In closed systems the temporal inverses of certain processes would require the spontaneous concatenation of a set of coordinated initial conditions.

For example, the isotropic contraction of a circular water wave to a point would require the spontaneous formation of a highly structured set of initial conditions. (Frankly, the exclusion of the water wave seems to us to be no more de facto than the reconstituting of the cigarette.)

Processes which are de facto irreversible are important to philosophical discussions of "time's arrow" because they seem to be counter-examples to the common opinion that entropy changes are the only reliable indicator of time's arrow. By categorizing a class of processes as de facto irreversible, one introduces a nonentropic but probabilistic criterion for irreversibility. Although governed in both directions by Newton's laws, the expanding water wave is overwhelmingly more probable than its inverse, and the concept of entropy is alleged to be irrelevant here. The exclusion of the inverse process is not ruled out nomologically. Even if there is an entropy effect due to dissipation, there is an independent sufficient condition for the irreversibility of the process, viz., the well-nigh impossible spontaneous concatenation of initial conditions which would yield the coherent contraction of the water wave to a point in space.

According to Grünbaum (1973, pp.267-269) no entropy is definable for processes of external expansion. Such processes in infinite open systems are offered as examples of nonconditional, nonentropic, de facto irreversible processes. While appreciating the idea of de facto irreversibility and the studied examples, we are of the opinion that a thermodynamic entropy can be defined for any system with macroscopic properties, including the expanding thin gas.

The distinction between de facto and nomological irreversibility is dubious; the idea of law-like irreversibility seems to be internally inconsistent. In the absence of any clear demonstration of a physical law which is temporally asymmetric, we must take law-like to imply reversibility. All fundamental equations seem to be invariant under time reversal and the essence of nomological irreversibility seems to require some counterinstance. Solutions to the fundamental equations may indeed reflect an asymmetry but that is inherited from the initial and/or boundary conditions which must be imposed to obtain a solution. Thus, the distinction between nomological and de facto irreversibility, as developed so far, seems to collapse; the irreversibility and asymmetry, in this approach, can only come from contingencies of initial or boundary conditions.

We will not enter here into a detailed critique of the notion of _de facto_ irreversibility and the examples given to illustrate _and_ validate. (For a move in that direction, see Zenzen, 1977.) Our problem here is to explain why processes that are not excluded _a priori_ are nevertheless excluded in natural phenomena. We cannot simply cite unfavorable initial conditions without explaining where they come from or leave the issue hanging on definitions of entropy. We have to find the historical cause of the initial conditions and the reason why that cause is signaled sometimes by entropy. The idea of _de facto_ irreversibility, apart from its comparison with nomological irreversibility, is very helpful because it suggests that there are different kinds of exclusion irreversibility. We will try to distinguish between _a priori_ and entropic irreversibility, and then wonder where statistical irreversibility fits in.

Entropic irreversibility refers to the exclusions of reverses of naturally occurring processes for which the reverses are conceptually possible, even at the molecular level, and yet somehow prevented. We have in mind especially processes governed by the second law of thermodynamics. The irreversibility is not _a priori_, nor is it generally _de facto_. Entropic irreversibility cannot be explained at the present time in terms of time anisotropy, nor can it be presented as the sole indicator of time anisotropy. The origin of the exclusion must be located somewhere in the underlying particle dynamics of the system behavior.

It may seem to be appropriate at this point to tackle head-on the problem of explaining entropic irreversibility without further attempts to sort out different kinds of irreversibility. There is, however, still a dark cloud hovering overhead, namely, statistical irreversibility. Most of the attempts to explain _entropic_ irreversibility during the past century have been aimed at origins of _statistical_ irreversibility, as if those two kinds of irreversibility were the same. Since we look for the origin of entropic irreversibility in mechanics, and not in _statistical mechanics_, we have to review the meaning of statistical irreversibility and then show where we depart from that.

5.7 Statistical Irreversibility: Necessarily de Facto

The argument here is divided into three parts. In Part (a) we develop a simple model which, in spite of its simplicity, illustrates some basic characteristics of statistical mechanics. Irreversibility does emerge in the description of even the simplest model, not as a contradiction to the mechanical reversibility, but as an indicator of the <u>reproducibility</u> of the phenomenon being described, and as a restriction on the kind of phenomenon which can be described.

Part (b) explores the time limits on the reliability of a statistical mechanical prediction, especially as they show up in the loss of periodicity that necessarily occurs in a statistical description. In Part (c) we calculate the statistical entropy of a simple model. This helps us to show how statistical descriptions fail to capture the full range of behavior allowed by the mechanical model. The statistical description "works" in the sense that what is observed is limited to cases that can be predicted statistically. This offers clues but does not explain why we do not see in nature excluded phenomena allowed by the underlying mechanical model.

What agreement there is between predicted and observed phenomena is seen to be contingent upon conditions which assure the reproducibility of the latter. Statistical prediction is validated by the reproducibility of the observed phenomenon. We want to understand why statistical descriptions "work," and we are led to ask why observed phenomena are reproducible (Note 8).

a) Probability and Statistical Entropy

In order to illustrate the properties of statistical irreversibility involved in later discussions, we will develop here a model for what seems to us to be an ultra-simple application of statistical mechanics. We consider first the case of a single particle moving in one dimension with a position denoted by x and a velocity v. There is no force acting on the particle. The set of possible states (x,v), i.e., the "state space," is the (x,v) plane, and a trajectory in that space under no force is a straight line, as illustrated in Fig.5.1a. Notice that the ordinate (v) is the velocity in the direction of the abscissa (x); that is the critical feature of "state space." The variety of possible trajectories is illustrated in Fig.5.1b where the arrow lengths represent velocities in the x-direction (distances covered in a common time t). The arrows show the "shearing motion" of the points in state space when the points move along trajectories.

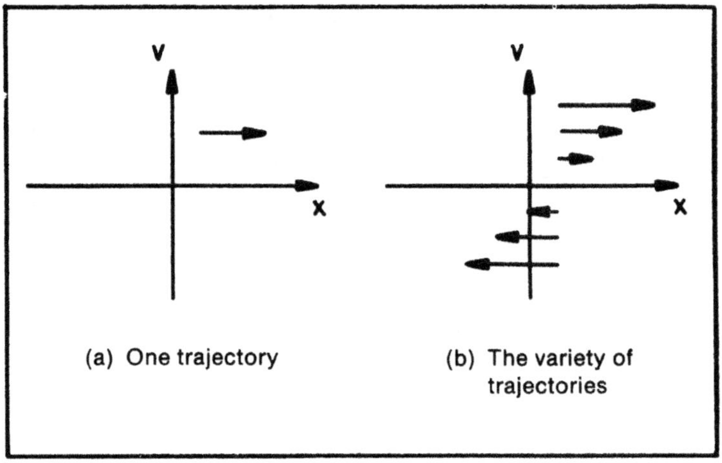

Fig.5.1 Trajectories in the State Space of One Particle

Now we consider the case of a particle with position x and velocity v known to within uncertainties Δx and Δv. Its initial state could be any point inside the rectangle B of Fig.5.2a. We will refer to the points (x,v) of the plane as the "microstates" and the rectangles as the "macrostates;" the intervals Δx and Δv are the "macroscopic data" which determine a macrostate. We will say that a microstate is "accessible" to a macrostate if it lies within the rectangle of the macrostate. There is a many-to-one correspondence between accessible microstates and macrostates; that is the essential feature of the model which allows it to be used to develop probabilistic predictions.

For a given macrostate, the statistical entropy is

$$S = k \ln \Omega \tag{5.7}$$

where Ω is the measure of the set of accessible microstates (Note 9). In the present case the measure Ω is simply the

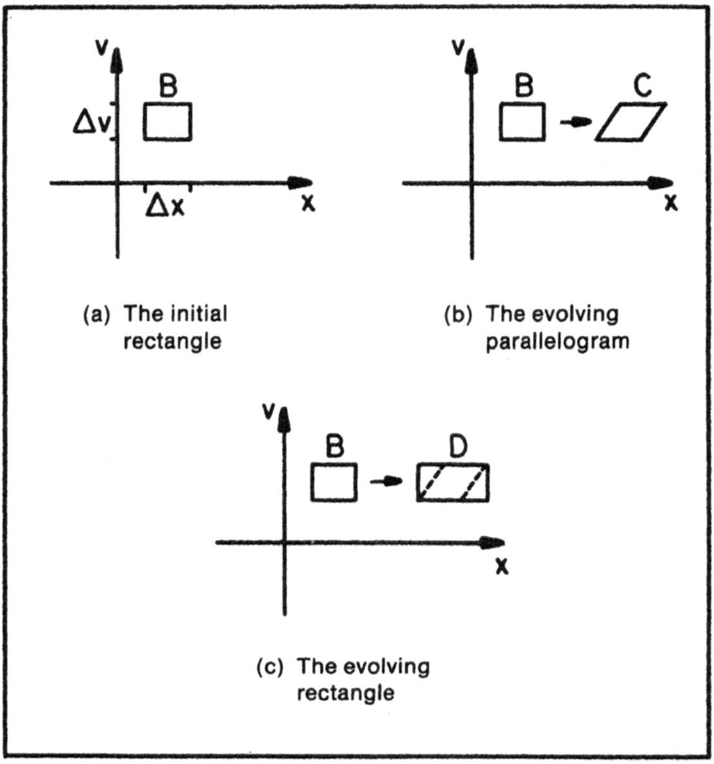

Fig.5.2 Accessible Sets of Microstates

area $A = \Delta x \Delta v$ of the rectangle (Note 10). We do not have to approximate Ω as a number W of microstates accessible to a "most probable distribution," as is common in connection with more complex models, leading to familiar expressions illustrated by

$$S = -k \int dx \int dv\, f \ln f \quad \text{or} \quad S = -k \sum_i n_i \ln(n_i/n) \qquad (5.8)$$

which go back, respectively, to Boltzmann (1872) and Jeans (1925). These expressions can agree asymptotically (in the limit of large numbers) with the more general Gibbsian idea expressed by (5.7) if the distributions described by f and n_i are the most probable of the accessible distributions (have the largest number of accessible microstates). The general formula (5.7) is sometimes expressed in a more familiar way

$$S = -k \sum P_i \ln P_i = -k\,\Omega\left(\frac{1}{\Omega} \ln \frac{1}{\Omega}\right) \tag{5.9}$$

where $P_i = 1/\Omega$ is the probability that the whole system is in one of its equally likely accessible microstates (Tolman, 1938, p.535 or Morse, 1964, p.252). In any case, the statistical entropy for our model is simply

$$S = k \ln A \,. \tag{5.10}$$

Let us now follow the evolution of the macrostate which is initially the rectangle B in Fig.5.2a. At a later time t the points inside the rectangle will have moved to form the set inside the parallelogram C in Fig.5.2b. It is easy to see that the area A_c of the evolving parallelogram is constant in time, which illustrates the Liouville theorem for the "conservation of measure" in state space (Tolman, 1938, pp.48-52). The problem that arises here is how to interpret the parallelogram C. It is not a macrostate (rectangle) although it has evolved from a macrostate. If we wish to describe the evolution of macrostates with this model we must either generalize the definition of macrostates or find a way to derive the macrostate corresponding to the evolving parallelogram.

In general, a macrostate is determined by macroscopic data, which are, in the present case, Δx and Δv. These properties are assumed to be independently measurable at any time, and the corresponding set of equally likely accessible states is the rectangle "framed" by these data. It is true, as we see in the evolution of rectangles to parallelograms, that correlations between x and Δv, or v and Δx do form in time, but a macroscopic description based on Δx and Δv is possible only to the extent that such correlations can be ignored. We have to define the predicted macrostate at time t by using the evolving parallelogram C in Fig.5.2b to identify evolving values for Δx and Δv and then using those, in turn, to frame the macrostate D in Fig.5.2c. The process of framing involves a loss of information but that is an inherent feature of our probabilistic prediction. The same kind of loss occurs in more complex models where framing becomes what is known as "coarse graining" (Tolman, 1939, pp.167-179).

According to the probabilistic prediction of the evolution of macrostates from the evolution of microstates, rectangle B in Fig.5.2c evolves to rectangle D. The area A necessarily increases, and hence the entropy increases. We could, of course, distinguish between two different kinds of entropy, (1) the "historical" entropy determined by the evolving parallelogram with no framing, and (2) the "dynamic" or "current" entropy determined by the evolving rectangle with framing. The historical entropy must remain constant in time, even in more complex models, because it measures only the microstates which have evolved from the initially accessible microstates. The dynamic entropy must grow because its measure includes additional microstates that become accessible relative to current values of macroscopic properties. This behavior is described very well in a general way by Jaynes (1965).

A shrinkage of dynamical entropy would be temporarily possible if the initial macrostate could be a parallelogram skewed in a manner opposite to the parallelogram C in Fig.5.2b, but that is not possible in the present case because the initial macrostate does not involve correlations between x and v. In any case, the dynamic entropy must reach a stage of irreversible growth and it must do this in either direction of time. The irreversibility is *a priori* irreversibility in the sense that its origin has nothing to do with the nomological character of the particle dynamics. It is due entirely to the probabilistic nature of the prediction as illustrated by the framing process.

One could argue that "coarse-graining," illustrated here in its most primitive form (framing), is not a necessary component of a probabilistic prediction. Granted that *a priori* irreversibility must show up in the behavior of a coarse-grained entropy, it is, nevertheless, possible to predict the evolution of the macrostate without referring to any entropy. Furthermore, probabilistic predictions are used in many applications that do not have an identified entropy.

We cannot answer the argument in general, of course. It may be true that all probabilistic predictions have a many-to-one correspondence between "states" in one description and "states" in another, which can be measured by an "entropy," but many applications do not have a conservation of measure for the evolution of accessible states. For the applications of statistical mechanics, assuming our model is illustrative, we can argue as follows.

Once the properties of the macroscopic description have been identified, in our case, Δx and Δv, the accessible microstates "become" equally likely. If the equal likelihood of microstates accessible to later intervals Δx and Δv is to be altered to reveal the accrued prejudice for values of x near the middle of the evolving parallelogram, then we have to (1) wonder about the initial rectangle, (2) replace originally selected macroscopic data by new kinds of data that will describe the prejudice, and (3) sympathize with later observers. The later observers cannot proceed as we did assuming that initial data determine the evolution. They must recognize the history of their system. They must be "psychiatric" in their approach.

If we accept the inevitable growth of entropy in a statistical mechanical model, illustrated here for a simple model and illustrated for more complex models by references cited above and below, even then we have to admit that this still does not explain the entropic irreversibility of observed phenomena. We have reason to believe that the irreversibility of statistical entropy is a priori; the entropy would grow in either direction of time if predicted properly. And we have reason to believe that observed entropy, faithful to the time reversibility of all phenomena described by our mechanical model, increases in one direction of time and decreases in the other direction. The conflict between these entropies can be viewed (1) as a paradox, or (2) as a clue to the origin of irreversibility in natural phenomena. The paradox is well known. The clue can be described as follows.

Suppose we have initial data for a macroscopic state and a molecular model for the system. Then even if we have no intention of developing a statistical prediction we can still identify the set of initially accessible microstates. What happens to this set as time passes? Does it "hang together" in the sense of remaining accessible to a single evolving macrostate, or does it separate into parts that are accessible to different macrostates? If it hangs together, then the evolution of the macrostate can be expected to be reproducible. Different microstates are the actual (and unknown) microstates in different trials, but since they all look alike at the macroscopic level, the same macroscopic evolution occurs in all trials. (By "all" here we mean, of course, very nearly all.) On the other hand, if the initially accessible microstates drift apart in the sense of becoming accessible to a growing variety of macrostates, then the macroscopic evolution would not be reproducible; it would go different ways in different trials. We would say the macroscopic evolution is chaotic and unpredictable. The requirement for reproducibility and predictability is that the initially accessible microstates (all but a negligible minority of them) remain accessible to the evolving macrostate, and this implies a nondecreasing entropy (measure of the set of currently accessible microstates). Decreasing entropy (for our isolated system) is a sure sign of chaotic behavior. The exclusion of such processes by the second law of thermodynamics is supported by the mere reproducibility of observed phenomena without any further appeal to statistical mechanics. Increasing entropy is a sign of assurance for reproducibility; the evolution not only remains "tight" in the sense that initially accessible microstates hang together, but it even becomes more and more like evolutions that could have proceeded from different initial conditions. New microstates are continually coming into the set of currently accessible microstates. This is what Boltzmann means when he describes the allowed macroscopic evolutions as ones that proceed from "less probable" (fewer accessible macrostates) to "more probable" (more accessible microstates) macrostates. In fact, Boltzmann's arguments about irreversibility are centered around the necessity of a trend from the less probable to the more probable (Note 11).

The clue that comes to us from comparison of statistical entropy and the entropy of observed processes is that we should focus our attention on the reproducibility of the observed processes. Because they are reproducible, as signaled by their increasing entropy if by nothing else, they are candidates for statistical descriptions. It follows that their reverses, with decreasing entropy, are not candidates, not reproducible. The reverses, unknown though they may be, must be chaotic and unpredictable. Perhaps we can substitute the question of where reproducibility comes from for the question of where irreversibility comes from. Before we explore that possibility we consider another paradox hoping that it may offer another clue.

b) Probability and Periodicity

There is a second paradox associated with the conflict between statistical and purely mechanical predictions about the evolutions of mechanical systems (Note 12). On the one hand, it is now well known (terHaar, 1954), having been discovered by Poincaré in 1890, that a Newtonian system of any number of particles, with or without interactions, is periodic if the system is confined by fixed external forces and isolated from changing external forces. On the other hand, it is also well known (Boltzmann, 1872, 1964; Ehrenfest, 1959; terHaar, 1954) that probabilistic predictions based on the Newtonian model decay in a way that rules out periodicity, and that macroscopic phenomena seem to have the Boltzmann decay rather than the Newtonian periodicity. The Newtonian periodicity is inherited by quantum mechanics (terHaar, 1954), so there is no escape from the paradox in the direction of quantum mechanics. On the other hand, the Newtonian period for an observable fluid can be older than the universe, so there might be an escape there. Furthermore, the periodic return to an initial condition does not have to be exact in the sense that a trajectory in the many-particle state space returns to its initial position in a finite time. The return is to "near" the initial position. Of course, "near" at the microscopic level means "exact" at the macroscopic level, so the recurrence is exact for macroscopic evolutions and it is puzzling that the macroscopic description contains no hint of periodicity.

Before we try to determine why periodicity is missing in macroscopic descriptions, let us consider what happens to periodicity in a probabilistic prediction. To this end we modify the model described above in two ways. First we impose reflecting walls to confine the particle to a fixed interval of length L on the x-axis. Then we will add other particles to the system.

The trajectory of a particle reflected by walls at $x=0$ and $x=L$ is illustrated in Fig.5.3a. The evolution of the rectangle B in Fig.5.2a under the new constraint is illustrated in Fig.5.3b, where the evolving parallelogram has advanced far enough so that part of it has been reflected. As time goes on, the parallelogram becomes more and more elongated until it becomes many strands, as illustrated in

Fig.5.3d. At such a time the location of the particle within the "box" has been entirely lost and the predicted macroscopic state is a rectangle that forevermore fills the box. Decay is evident.

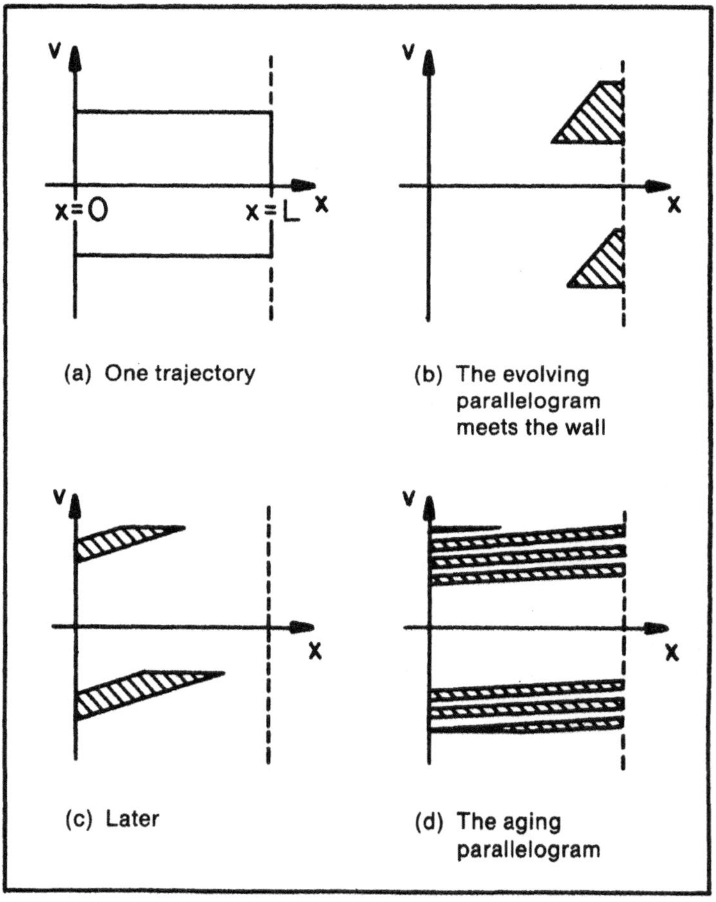

Fig.5.3 Evolutions with Reflecting Walls

Now suppose that we start with a beam of many particles which are represented by the dots in Fig.5.4a. The density profile n(x,t) at the initial time t = 0 is shown in Fig.5.5a. As time passes the evolving parallelogram elongates, as before, with the particle dots embedded in it, as illustrated in Figs.5.4b and 5.5b. The derivation of Fig.5.5b from Fig.5.4b should be quite reliable, for some time. Eventually, however, the configuration of the particles can return to very nearly the initial configuration even though the evolving parallelogram continues to decay (elongate) as illustrated in Fig.5.4d. The predicted density profile in Fig.5.5d is totally unreliable. The observed profile would be the one shown in Fig.5.5a. There is nothing very surprising in these evolutions. The point is that a probabilistic prediction must decay and become less and less reliable until it has lost all potential for predicting departures from uniformity.

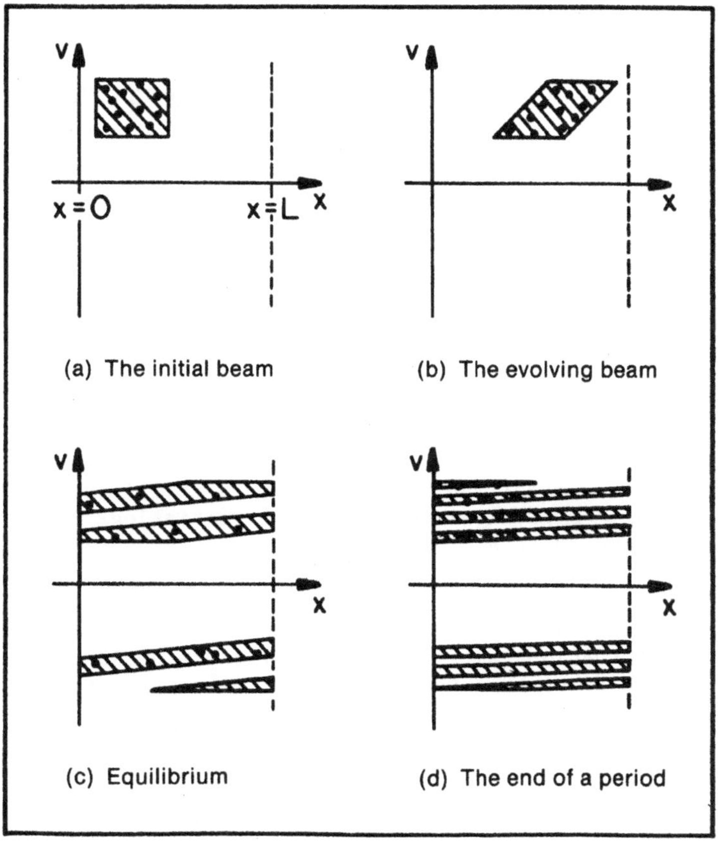

Fig.5.4 Hiding Periodicity: The Decay of the Probability Fluid

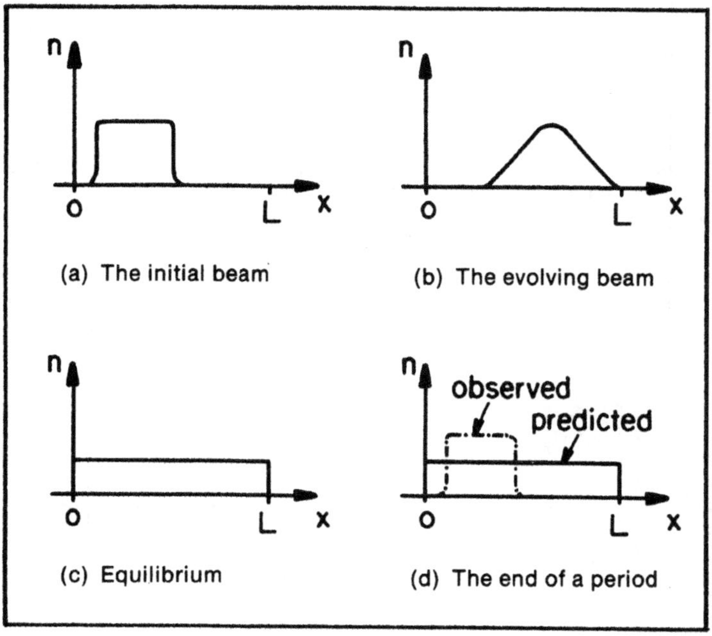

Fig.5.5 Density Profiles for the Beam of Fig.5.4

c) Probability and H-Theorems

The beam model was introduced above to illustrate what happens to mechanical periodicity in a statistical description interpreted as a predicted feature of observed phenomena. But beyond that it also provides a second example of the nature of the increase of statistical entropy for Newtonian models, and that example comes closer than the first to situations of interest in later discussions.

Notice first that the statistical entropy of the beam model is not defined by Eq.(5.10) even though it is, of course, still defined by Eq.(5.7), where Ω is the measure of the set of accessible states in the many-particle state space. The exact definition of statistical entropy for this model, as for any model, depends on the selection of the set of accessible microstates, and that, in turn, depends on the selection of macroscopic parameters. Once those parameters have been identified and the accessible microstates found, then the microstates can be followed in time to predict the evolution of the macroscopic parameters, and the values of those at any time can be used to rediscover, i.e., to frame, the currently accessible microstates and to evaluate the dynamical statistical entropy.

Unfortunately, the framing process in the many-particle state space is difficult to visualize or to carry out explicitly. The literature is full of references to the "coarse-grained" entropy of statistical mechanics, but a precise visualization of the coarse-grained process is rare; it can be developed only for models as simple as the one-particle model described above. Even the beam model, which is only slightly more complicated in many respects, is beyond framing.

Fortunately there is a well-known way to approximate the statistical entropy for many models, and the ensemble methods of Gibbs extend this approach to any model of thermodynamic interest. The approximation becomes exact in the limit of large numbers of particles provided it is applied correctly. The method involves distributions of particles or other subsystems over their states. Corresponding to the macroscopic data there may be many accessible distributions, and corresponding to each distribution there may be many accessible microstates. If the most probable distribution is overwhelmingly most probable in the sense of claiming the overwhelming majority of accessible microstates, then the entropy of the macrostate can be approximated as the entropy of that most probable distribution which is given by familiar formulas such as the one in Eq.(5.8). (For a deep analysis of the form of formulas for entropy, see Grad, 1961.)

Let us try to calculate the entropy for the beam model from Eq.(5.8). If we regard the evolving density f(x,v,t) of particles on the one-particle state space as a most probable distribution relative to macroscopic data, then the entropy is the H-function entropy of Eq.(5.8). It can be shown by following the evolving parallelogram of Fig.5.4b or by a more general argument (Zenzen, 1977) that such an entropy for noninteracting particles remains constant in time. This conclusion has led, as we will notice below, to the idea that the creation of statistical entropy requires intermolecular forces or collisions. Otherwise the entropy increase due to the spatial expansion of the gas can be compensated by the entropy decrease due to "velocity sorting" (larger velocities more prevalent at greater distances) (Zenzen, 1977). The idea is correct if we assume that the H-function entropy is the appropriate entropy, but that, in turn, has to be tested against the macroscopic data.

In general, even if the one-particle distribution function f is determined initially by the macroscopic data, the H-function entropy is not appropriate. It is appropriate initially, when f is the most probable f, but that f evolves into an f which does not remain most probable. In fact, it is well known that conductive transport phenomena are due to the difference between the evolved f and the local equilibrium (most probable) f (Note 13).

Where does all of that leave us in our pursuit of the statistical entropy for the beam model? We cannot use the simple framing process used above in connection with the one-particle model, which is too "coarse" in the sense that it ignores the macroscopic density profile n(x,t) plotted in Fig.5.5b), and we cannot use the H-function entropy because it is too "fine" for noninteracting particles. Let us use the macroscopic function n(x,t) and the limits on the velocities v. (In more complex models we encounter more macroscopic data, such as temperature and various densities, but here we have only a little more macro-information than we had in the one-particle model.) We identify the most probable distribution as follows:

$$f(x,v,t) = \begin{cases} 0 \text{ if } v \text{ not in } \Delta v \\ (1/\Delta v)n(x,t) \text{ if } v \text{ in } \Delta v \end{cases} \tag{5.11}$$

which satisfies the requirement of Fig.5.5b and the constraint

$$\int f\,dv = n(x,t)\ . \tag{5.12}$$

This formula could be derived by the method of Lagrange multipliers, which is certainly the method to be used in more general cases, but the explicit derivation here is trivial. Using Eq.(5.11) in Eq.(5.8) we find

$$S(t) = -k \int dx\, n\, \ell n\, n\ . \tag{5.13}$$

The rate of change with time is

$$dS/dt = -k \int dx(\partial n/\partial t)\, \ell n\, n + (d/dt) \int dx\, n\,. \qquad (5.14)$$

The second term on the right side of Eq.(5.14) has zero value because the integral of n is the constant total number N of particles. The first term can be simplified (replacing the time-derivative by an x-derivative) by using the "equation of continuity" (conservation of mass), which always holds, even for the simplest models,

$$(\partial n/\partial t) + (\partial/\partial x)\, n\bar{v} = 0 \qquad (5.15)$$

where $\bar{v}(x,t)$ is the average velocity, which can be visualized from Fig.5.4b and which can be calculated (no need) from n(x,t) and initial data. The only feature of the derived function $\bar{v}$ that must be noticed here is that it is positive on some x-intervals within the range of the parallelogram in Fig.5.4b and zero elsewhere. Now we have Eq.(5.14) as

$$dS/dt = k \int dx\left(\frac{\partial}{\partial x}\, n\bar{v}\right)\ell n\, n\,. \qquad (5.16)$$

Integrating repeatedly by parts and using the zero-value boundary values, we have

$$dS/dt = -k \int dx\, \bar{v}\, \frac{\partial n}{\partial x} = k \int dx\, n\, \frac{\partial \bar{v}}{\partial x}\,. \qquad (5.17)$$

Since both factors of the integrand on the right side are positive on some x-intervals and zero-valued on other intervals, the rate of change of the entropy is necessarily positive.

This conclusion of increasing entropy for the beam model can be extended to longer times to allow for the particle-wall collisions after the further elongation of the parallelogram in Fig.5.4b, and it can be extended to the case of any collisionless model for which the macroscopic parameters can be related to moments of the one-particle distribution function (Note 14). Even more generally it has been shown in many ways that the coarse-grained entropy of complex models increases; this is one of the areas in which the literature of statistical mechanics has been very eloquent.

Our purpose has been to suggest, perhaps hastily, in view of the simplicity of our models, that statistical predictions of macroscopic phenomena on the basis of molecular models have certain deficiencies. They are necessarily irreversible in the sense of criterion 3 even if the processes they describe are reversible in the sense of 2 and potentially reversible in the sense of 3. When there is agreement between the statistical description and observed phenomena then that agreement is contingent upon conditions for the latter which assure the reproducibility of the phenomena. Furthermore, the statistical predictions fizzle too soon to show the periodicity of the underlying mechanical model. If we persist in accepting the underlying mechanical model, then the absence of its periodicity in observable phenomena must be contingent upon natural conditions which affect the observable system. The suppression of periodicity in statistical descriptions, and the inevitable irreversibility of such descriptions seem to show up in observable phenomena, but the agreement can only be a clue to help identify the natural origin of irreversibility and decay. The mechanical model allows for a wider display of behavior and we have to ask why the observed display is limited to cases that can be predicted, and predicted statistically. Why are the deficiencies of the statistical description so useful?

CHAPTER 6

PROPOSED ORIGINS OF IRREVERSIBILITY

The earliest approach to the resolution of the paradox of how macroscopic irreversibility can emerge from a reversible molecular model was to question the model itself, or the methods applied to develop the model. (We refer here back to Chapter 2.) It is difficult for us today to appreciate the problems of those days, for we all believe today in atoms and molecules and in some kind of dynamical law. We have to try to put ourselves in the turmoil that attended the final stages of the birth of the molecular hypothesis. Fortunately, we have excellent studies, (Brush, 1967) including the one by Daub (1969), which should be read after the study of earlier problems by Koenig (1959). And we have the heavy major contributions referred to in Chapter 2.

In retrospect we may say that part of the difficulties encountered in those early arguments was due to the failure to distinguish among various kinds of irreversibility. The word "irreversibility" seemed to have a single referent, and that referent had to be more versatile than is logically possible. In fact we still have that problem today; if we try to sort out different kinds of irreversibility we wonder whether we are engaged in semantics or mired in paradoxes.

Going back to the beginning we notice that when Boltzmann found (1872) an irreversible statistical description based on the reversible Newtonian mechanics, he was challenged severely for thirty years by thinkers who argued that the model could not be valid because there is no way for macroscopic irreversibility (criterion 2) to arise from molecular reversibility, and there is no way for fluid decay and dissipation to emerge from the periodicity of particle dynamics (Boltzmann, 1964; Ehrenfest, 1959; terHaar, 1954; Daub, 1969).

Boltzmann emphasized, in subtle ways, that the irreversibility in his description arose because of the probabilistic nature of the description, not because of any violation of the underlying Newtonian mechanics, as if to anticipate the later distinction between <u>nomological</u> irreversibility (i.e., law-like, by natural law) and <u>de facto</u> irreversibility (arising from the form of the description). It was no surprise to him that reversibility and periodicity could be destroyed by a probabilistic element in the description.

What was surprising, and still is, was that such a description seems to fit observable phenomena. This is not to say that the origin of irreversibility in natural phenomena is probabilistic, but rather that the success of the statistical descriptions raises new questions, offers new clues. In fact, it did raise, for the first time, the question of how irreversibility, decay, approach to equilibrium, and all of those characteristics of macroscopic fluid behavior, can be reconciled with the periodicity, the reversibility, the persistent motion of the molecular model.

It is, in retrospect, unfortunate that Boltzmann did not stress the distinction between the probabilistic origin of *a priori* statistical irreversibility and other nonprobabilistic origins for irreversibility in natural phenomena. He was preoccupied with the marvel of the agreement between statistical and observable descriptions and with the evident trend in nature toward more probable states. His opponents (Ehrenfest, 1959 and terHaar, 1954), we believe, argued correctly against the notion that some intrinsic natural probability could be the basis for natural irreversibility. His early defenders (Ehrenfest, 1959 and terHaar, 1954) also misunderstood his arguments, so much so that they found their way into arguments for discontinuous jumps in the H-function (at each molecular collision) and arguments for very large probabilities of entropy increases at each jump, as if *a priori* irreversibility could be patched up by physical reasoning to make it less *a priori* and more explanatory, forgetting the continuity of probability densities and the continuities dictated by differential or integro-differential equations.

Now we have a legacy of arguments proposing corrected or refined probabilistic origins for nature's irreversibility. We even have a separation of time scales between scales in which the change is "smoothed" down to reach a "hydrodynamic stage" (Bogoliubov, 1961).

6.1 Probabilistic Origins

The arguments for a probabilistic origin of observed entropic irreversibility are of three distinct types:

(a) Irreversibility comes from stochastic elements that interrupt the otherwise deterministic lives of individual particles.

(b) Irreversibility comes from hidden constraints or hidden variables that leave in their "wake" room for uncertainty and irreversibility.

(c) Irreversibility comes from the statistical method used to describe the evolution of the system.

We will discuss each of these.

a) Underlying Stochastic Elements

According to arguments of this type there is an unknown, non-Newtonian, non-Schroedinger, random influence that enters the lives of individual particles to make them forget where they were going. The uncertainty of their motion is to be added to the uncertainty about initial data for starting points of trajectories. It is assumed that there are uncertainties not only in the initial data for individual particles but also in the underlying particle dynamics. The assumption is more severe than the use of "transition probabilities" in quantum mechanics, where the scheme is deterministic but expressed in terms of predicted transition probabilities. The assumption is more severe than the adoption of stochastic elements for data-fitting purposes in admittedly empirical descriptions. In fact, the assumption is tantamount to a denial of the original mechanical model, and it is not clear what purpose it can serve other than refuting the model.

Proposals of truly stochastic interruptions are rare. The common reference to Kac (1959, p.86) is hardly justified in view of his earlier writings and later interpretations (Wannier, 1966, pp.396-403). On the other hand, there is a common practice of referring to a "loss of memory" during the trajectory of a particle or an aggregate of particles followed mechanically or statistically (Note 15). Sometimes the expression, "loss of memory," is intended to describe some aspect of a priori statistical irreversibility, which could be described better as a "loss of information." We believe that terms which suggest denial of the underlying mechanical model should be avoided unless they are to be taken literally.

And there is a new somewhat mystical proposal for our consideration coming from one who has contributed so many advances to our field, namely, I. Prigogine. He now refers to the need for a revision of the underlying molecular dynamics to allow for an "intrinsic" stochastic nature at the molecular level. He sees the paradox of irreversibility as serious as the challenges to molecular mechanics presented by relativity and quantum mechanics, and sees the kind of need for a revision that was fostered by those innovations (Prigogine, 1980,1983).

b) Hidden Constraints and Hidden Variables

Arguments of the second type proposing hidden constraints, etc., go back, at least, to E. Einstein (1922). It is instructive here to notice what she tried to do. It was well known that the density, velocity, and temperature profiles of a pure, dilute gas are sufficient data to determine reproducible values for transport coefficients for heat conductivity, momentum conductivity (viscosity), and self-diffusion (Jeans, 1954). The first step of the established theory was to identify "constraints" (independent variables) and to find the most probable distribution of molecular velocities which could then be used in the second step to predict transport rates for a gas not too far from thermal equilibrium. Einstein suggested that there might be another constraint (independent variable) and chose the thermal conductivity as the candidate. She used thermal conductivity as a constraint and used the method of Lagrange multipliers to find the correspondingly most probable distribution of molecular velocities. Then she used that distribution to calculate the viscosity.

The results were puzzling. In retrospect we can say that they should have been puzzling. If the earlier choices of independent variables (constraints for determining accessible states and the most probable distribution of molecular velocities) were already successful, then there can be no way to add new constraints and still be successful. The purpose of the statistical description is to explain why the variables found to be dependent are actually dependent; assigning arbitrary values to one of these variables can only interfere with the development of a statistical description.

We encounter a question hinted at earlier: how do we know when we have enough macroscopic data, enough independent variables so that other variables can be predicted at any time and evolutions can be predicted uniquely (reproducibly)? This question is the same as the question of how do we know when we have enough, and not too much, macroscopic data for determining accessible microstates and calculating the appropriate entropy? We cannot answer this important question, except to point to the guidance of fluid dynamics, especially thermodynamics, and to notice the proofs of negligible fluctuations at equilibrium for the thermodynamic choices of independent variables. (See, for example, Hill, 1956, Chapter 2.) The search for hidden variables in statistical mechanics, and also in quantum mechanics, from which the hidden variable could filter to statistical mechanics, has not yet been successful in the sense of having an influence on current treatments of fluid dynamics.

There is a hint of a way to look for hidden independent variables in statistical mechanics. We refer to the "ergodic theory," due originally to Boltzmann and developed by others (terHaar, 1954, Farquhar, 1964). The question of ergodic theory is whether or not a particular trajectory in the many-particle state space, starting at an accessible state, moves around so widely as to spread itself throughout the set of accessible states (Note 16). If it does, then the time average of any macroscopic property along that trajectory agrees with the state space average over the equally likely accessible states, and that equivalence has been held up as a justification for trusting averages over the states. One may question the assumed primacy of the time average (Sklar, 1973), but the equivalence could offer clues if we could assume that the trajectory densely covers the accessible states. The trajectory does move around in such a way as to become "dense" in some subset of state space, but that subset could be, and usually is, only a "corner" of the whole set of accessible microstates. Whatever it is, the covered subset is determined by all of the constants of the motion, some of which correspond to macroscopic variables not yet recognized. These are the "hidden variables" of ergodic theory.

Khinchin (1949) refers to the subset determined by all possible constants of the motion as "metrically indecomposable" and suggests that a trajectory starting within such a subset does move around throughout the subset, but he admits that most sets of accessible microstates are decomposable. Farquhar argues that the question of the ergodicity of state space has no connection with the problem of explaining irreversibility (Garrido, 1972, article by Farquhar). Noticing these limitations of ergodic theory, and agreeing completely with others (Jaynes, 1965; Sklar, 1973; Strauss, 1972) on the irrelevance of the theory in relation to statistical mechanics, we ignore the possibility of hidden independent macroscopic variables as suggested by ergodic theory.

As for the hidden variables which have been considered as sources of mystery in both statistical and quantum mechanics, we have to agree with Jammer (1974) that it is not yet clear what purpose they could serve. There are arguments on the other side (de Broglie et al., 1976).

c) Coarse Graining and Time Smoothing

The third type of argument about the probabilistic origins of entropic irreversibility points to the statistical method used to describe the system. Here we encounter such a large literature that it is difficult to know where to begin. Probably no one has ever argued seriously that a method of description is the cause of any attribute of nature, including nature's irreversibility, but the literature of statistical mechanics is full of statements that seem to be aimed in that direction. Most of the arguments deal with *a priori* statistical irreversibility rather than the irreversibility of natural phenomena.

Let us begin with the Boltzmann H-theorem. As was mentioned above, there are reasons to believe that Boltzmann himself was aware of the *a priori* nature of statistical irreversibility, but he did not communicate this point well enough to prevent later attempts to generalize the H-theorem into a demonstration of the second law of thermodynamics. It is not surprising that so many attempts to prove H-theorems for more complex models have failed, or that early attempts to rationalize the H-theorem with the time reversibility (criterion 2) of the underlying particle dynamics (Ehrenfest, 1959 and terHaar , 1954) continue to be very difficult to understand. In retrospect we may say that the effort was too much for mere *a priori* statistical irreversibility. The authors were after bigger game, and they developed clues, but there was no way, from the beginning, to find nature's irreversibility in statistical descriptions. There was little progress in the development of the theory of irreversibility until 1940.

In the 1940's there were some remarkable efforts to generalize statistical mechanics to bring the nonequilibrium theory to the level provided by Gibbs for the equilibrium theory at the turn of the century (Note 17). One of the byproducts of this development was a new interest in the origin of irreversibility, and there arose a new language to supplement the Ehrenfest-Tolman references to "coarse graining."

It was suggested, for example, that irreversibility is an asymptotic feature of statistical descriptions which depends on going to the limit of large numbers N of particles or degrees of freedom and/or large volumes V, and sometimes large times t, reflecting, perhaps, the strange use of large times in older perturbation theories (Note 18). It is true, of course, that the macroscopic formulas of statistical thermodynamics are necessarily asymptotic (large N,V), but the reason for that has nothing to do with irreversibility. It is clear to us, in view of the previous section, that statistical irreversibility is not an asymptotic thing; it shows up quickly in the description of even just one particle.

The continuum infinity of points in a probability fluid on some part of a particle state space has no relation to the number N of particles in the physical system. The latter is related to the dimensionality of the state space, not to the population of the "ensemble." Carelessness in this area has led to a number of errors, even to the notion that the properties of a probability fluid (e.g., H function, n-particle distribution functions) jump discontinuously in time when particles collide.

Some of the concern about the number N of particles has been based on the idea that the Poincaré period (recurrence time) must be infinite in order to allow irreversibility to enter the statistical description, and that requires infinite N (Note 19). As we saw in the previous section, however, the Poincaré periodicity is necessarily suppressed in any statistical mechanical description; there is no need to introduce assumptions to kill the periodicity. The problem is not to explain why periodicity is missing in statistical descriptions but to explain why it is missing in observed fluid behavior, assuming that behavior can be accounted for in terms of a mechanical model.

It has been suggested that irreversibility in statistical descriptions requires interactions between particles which "drive" the system toward equilibrium (Note 20). It is easy to reject interactions, including "collisions," as the driving force for irreversibility or approaches to equilibrium because there are examples of irreversible approaches to equilibrium which are much faster when there are no interactions (Biolsi and Hollinger, 1974). In fact, this aspect of the nature of irreversibility is very familiar. Think of the diffusion of a colored gas from a broken ampule in an evacuated container (instantaneous) as compared to the diffusion in a container filled with air (takes hours). We sometimes think of diffusion as an irreversible process caused by collisions. On the contrary. It is an irreversible process, but it is inhibited and drastically slowed down by collisions. The common ideas and formulas for diffusion, and other transport phenomena, are restricted to cases where the inhibition is so severe that the fluxes have reached "terminal velocities" against the "friction" of collisions. Interactions in statistical descriptions, or even in mechanical descriptions, cannot be a cause of irreversibility.

Most of the attempts to identify the origin of irreversibility in statistical descriptions, often in the spirit of assuming that such an identification would also cover irreversibility in nature, have been centered around various forms of "coarse graining," starting with the original concept due to the Ehrenfests (Ehrenfest, 1959; terHaar, 1954; Tolman, 1938) and proceeding to many other forms including "time smoothing" (Note 21). We believe that some form of coarse graining does arise naturally in statistical mechanics, but some versions, especially those related to time smoothing or to any procedure imposed ad hoc without necessity, are irrelevant. They may help in a roundabout way to derive the Boltzmann equation from the Liouville equation, but they dissolve the possibility of extending the equation to three-body and higher correlations (Hollinger, 1962). The inherent coarse graining of statistical mechanics, e.g., the "framing" process described in the previous chapter, does explain the irreversibility of statistical descriptions, but not the irreversibility of natural phenomena.

Similar to coarse graining procedures is Boltzmann's assumption of "Stosszahlansatz" and its modern equivalent, "molecular chaos," sometimes to be repeated continually. More recently we have the "dynamics of correlations," which goes by many names, or just "randomness," or "random phasing," maybe repeated continually, and the "dissipative part" of perturbations (Note 22). We also have the "flow of ignorance," the "loss of information" or the "transfer of information" to inaccessible places within the machinery of the description, the "sacrifice of knowledge," and so forth.

The trouble with the new language and the new elaborations of old themes, from our point of view, is that they deal, at best, with the a priori irreversibility of statistical mechanics rather than the irreversibility of observed phenomena. They add more clues than we can digest, and they offer many important articulations of basic ideas in statistical mechanics, but they do not come directly to the problem of interest here.

6.2 Mechanical Origins

If we look for nonprobabilistic origins of entropic irreversibility we find many suggestions, some of which have been offered in connection with statistical arguments. It has been suggested, for example, as was mentioned above, that irreversibility comes from the large number of particles in the system of interest. If we now interpret that in a purely mechanical way the suggestion becomes more interesting and we are tempted to ask how large the number must be. We are no longer trying to derive asymptotic statistical formulas or destroy the Poincaré periodicity, so we do not expect infinity in the answer.

Another suggestion mentioned above is that irreversibility comes from interactions or collisions among particles. Here we add references to anharmonicities, retarded action, unnoticed perturbations, "impurities" (Note 23). Most of these "origins" of irreversibility are as sterile as any kinds of interactions but we should notice two possibilities. First, the retarded action of electromagnetic theory, and its counterpart in general relativity, not to mention the idea in its general form, identify a kind of exclusion irreversibility. In fact, the problem of deciding whether or not retarded action is an origin of irreversibility, when it seems to be and yet the equations are invariant under time reversal, is resolved by admitting that processes depending on retarded action are reversible in the sense of criterion 2, but irreversible in the sense of criterion 3. On the other hand, it seems very unlikely that the exclusion irreversibility of retarded action, which would take too long to manifest itself in ordinary observed fluids, could be the origin of the exclusion of heat going the wrong way. Some authors make a connection between the irreversibility of retarded action and a "fact-like" irreversibility that reminds us of the _de facto_ irreversibility described above (Section 4) (de Beauregard, 1970) and others have considered retarded action in their pursuit of cosmological origins of irreversibility (Gal-Or, 1976). But we are not yet compelled to focus on retarded action as the origin of entropic irreversibility.

The second possibility is in the area of unnoticed perturbations. Perturbations which are due to the interactions among particles within the system of interest have already been excluded from further consideration. But let us consider the accumulation of perturbations from outside influences. Certainly those perturbations are physically present, and they do interfere with the Poincaré periodicity of the system. The problem here is that entropic irreversibility is not a delicate property that can be caused by "perturbations." If we consider the larger macroscopic changes of external influences, not "perturbations" but drastic changes, and the quick and drastic responses of the system, then we come closer to a possible origin of entropic irreversibility.

It has been suggested that irreversibility has its origin in the quantum mechanical nature of physical systems (Born, 1948). And there is always the Uncertainty Principle to fall back on, or the idea of how measurements change the system, as if that were uniquely quantum mechanical. But there has been no development of these ideas toward our goal, and we cannot provide any. We also notice tendencies to argue against the quantum origin (Note 24).

It has been suggested that the origins of several kinds of exclusion irreversibility, including entropic irreversibility, are cosmological, coming somehow from the distant past, along with the expansion of the universe (Gold, 1967; Mirman, 1975; de Beauregard, 1970; Gal-Or, 1976). Without disagreeing with this suggestion we look for an everyday mechanism that expresses the bias of the past.

Our own views depend particularly on clues from the success of statistical descriptions and the idea of external perturbations. We will also focus attention on macroscopic equilibrium, not as a condition to be approached, which is the familiar way of viewing equilibrium in statistical arguments, but as a condition to be left behind. We view the derivation of fluid dynamics from molecular dynamics, whether by means of statistical mechanics or any other method, as a description of how macroscopic systems _leave_ equilibrium, not how they approach equilibrium. Naturally we hope that the macroscopic description of the _leaving_ will be good enough to persist through the passage to the next equilibrium, and experience tells us that our hopes are usually justified. But if we ever have to choose between studying _leaving_ and _approaching_, the choice will be as easy as it is for a rifleman who knows that his aim can never be better for distant targets than it is for close targets.

CHAPTER 7

THE ORIGIN OF EXCLUSION IRREVERSIBILITY

7.1 The Simplest Newtonian Models

Let us consider somewhat qualitatively the behavior of mechanical systems of increasing complexity, looking for hints of the irreversibility defined by criterion 3. First we consider a particle moving along a line (the x-axis) with position x(t) at time t and velocity v(t) under the influence of a force which does not vary in time and which restricts the particle to a finite interval. For example, this particle could be bouncing back and forth between rigid walls or it could be oscillating at the end of a spring. A graph showing the variation of x(t) would look something like the graph in Fig.7.1a. Let us refer to that kind of variation as changing and "periodic." We could just as well be following a particle in a three-dimensional box and following any of its coordinates.

Consider next a collection of particles, few or many, and let them collide or not. If we follow the position of one particle in time we would find a variation like the one illustrated in Fig.7.1a (no collisions with other particles) or the one illustrated in Fig.7.1b (with collisions). The variation is periodic in either case, but a single period in the second case may include many "fluctuations;" it is not only periodic but also "noisy" or "erratic."

Now follow the center of mass of the collection of particles, two or more particles, better if ten or more, even better if a thousand or more, and still better if allowances are made for the temperature of the wall molecules (Morrissey, 1970). We could just as well follow the velocity of the center of mass or a count of the number of particles on some interval or the added kinetic energies of those particles (temperature) or any other property P of a "collective nature" which depends on the positions and/or velocities of the particles on some interval. The variation with time of such a collective property turns out to have a pattern like the one illustrated in Fig.7.1c. The graph there is sketched to be similar to the graph in Fig.7.1b but it shows a feature not evident in Fig.7.1b that will become the key element in our proposed mechanism for irreversibility. It shows the beginnings of the occurrence of "equilibrium plateaus."

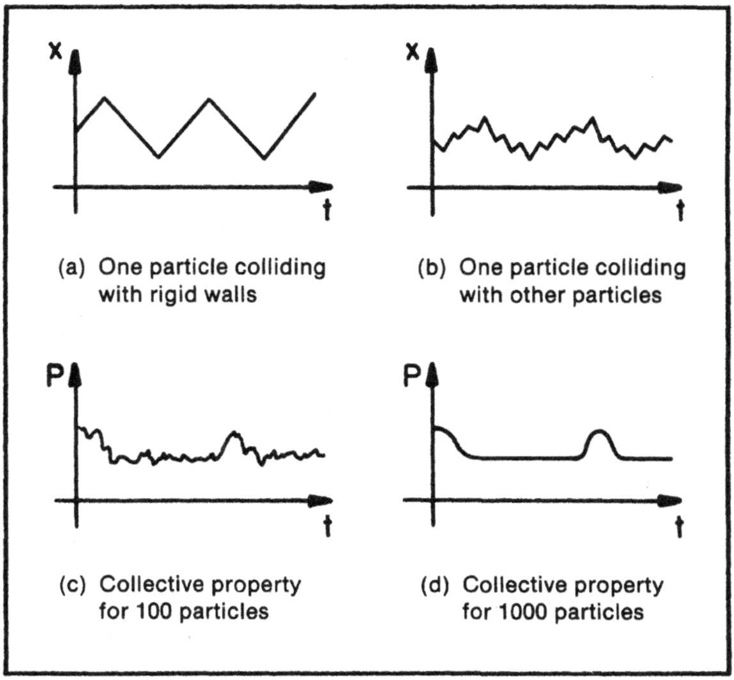

Fig.7.1 Patterns of Change: Emergence of Equilibrium Plateaus

The equilibrium plateaus are barely discernible for systems of a small number of particles, but they are very clear for fifty or a thousand particles in any reasonable "thought experiment," or any such experiment carried out on a computer (Morrissey, 1970). By the time we reach a thousand particles, even if they have no collisions, the variation of collective properties, suitably scaled, can look like the variation illustrated in Fig.7.1d. We will refer to that pattern of change as one with equilibrium plateaus interrupted briefly by repetitions of the initial withdrawal from equilibrium.

The situation we have begun to consider is illustrated in more detail elsewhere (Hollinger, 1972). Consider Fig.7.2a, where the collective property P is plotted against time, showing an initial forced withdrawal at time t_0 followed by a relaxation to equilibrium, reaching equilibrium at t_1. The "forced" withdrawal at t_0 could be, for example, the expansion that follows an outward displacement of a containing wall if the gas has spent a short enough time at a previous equilibrium to be on the "young part" of an equilibrium plateau (i.e., to be assuredly "randomized" at t_0 and long thereafter). Then the mechanical model for the gas will relax to an equilibrium plateau, which starts at t_1.

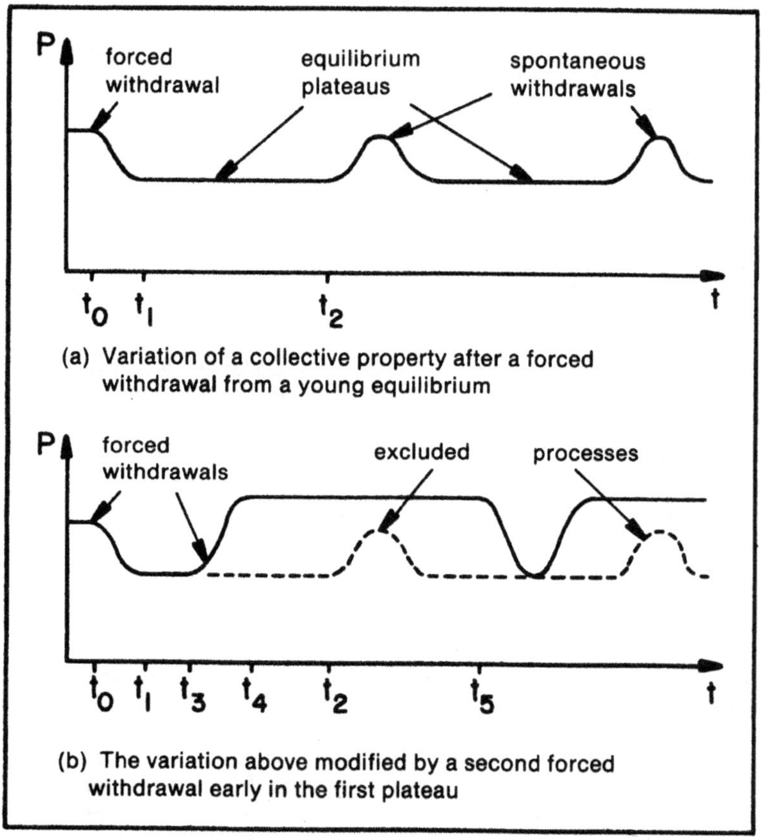

Fig.7.2 A Mechanism for Exclusion Irreversibility

The relaxation process is "young" in the sense of having followed a recent forced withdrawal and being near the beginning of an equilibrium plateau. The plateau itself will go on and on for a time that becomes very long, compared to the time of the relaxation, if many particles are involved. The most important feature of fluid behavior that comes from the presence of so many molecules is the long length of almost all plateaus.

Eventually, if the mechanical model is reliable, and we assume it is, there will be macroscopic fluctuations away from the plateau, even if the system has been isolated all the while, and some of these fluctuations will be returns to the initial conditions at t_0. Such returns are "old processes," occurring at ends of plateaus, without any external causes. We may call them "spontaneous withdrawals" from equilibrium, using the term "spontaneous" literally and ignoring the more common usage which identifies forced withdrawals. The spontaneous withdrawals are periodic variations which are guaranteed by the periodicity of the mechanical model. Such a withdrawal is followed by another period dominated by the equilibrium plateau and another withdrawal, etc. In Fig.7.2a the spontaneous "old" withdrawal starts at time t_2 and is followed by a repeat of the "young" forced withdrawal which started at t_0. There is a succession of young processes followed by plateaus followed by old processes.

These features of the variation of a collective property for a few particles (short plateaus) or more particles (longer plateaus) can be illustrated easily by thought experiments, fun experiments (people walking across a room), computer experiments, especially for the case where there are no collisions, and that is the most challenging case for our purpose here. (The tendency to form and preserve equilibrium plateaus is strengthened by interactions.) It is found, as mentioned above, that the larger the number of particles, the longer the plateaus. The duration of a noticeable fluctuation also increases with the number of particles, but not nearly so fast as the lengths of equilibrium plateaus. (The possibility of fluctuations that are not returns to initial conditions decreases with increasing numbers of particles.) We will notice the same behavior in quantum systems in Chapter 10.

It is easy at this point to stimulate skepticism about the on-going argument because of the simplistic nature of the model and the absence of a background in classical physics for such arguments. It is true that there is nothing very sophisticated about collisionless gases and no theory about equilibrium plateaus, but that is just a historical accident. There was never an obvious need for the science of equilibrium plateaus in many-body systems. It may be that such a science should be developed.

Returning to our model as described by Fig.7.2a, we now ask what happens if there is a change of external influences at a time t_3 between the beginning t_1 and the end t_2 of the equilibrium plateau? Then the system is forced to leave the plateau, as indicated in Fig.7.2b. This forced withdrawal leads to a new plateau (all such variations lead to plateaus) starting at t_4 and modifies the history of the system so that it will not undergo the withdrawal at t_2. That spontaneous withdrawal is replaced by another which would occur at some time t_5. And if there is another interruption (change of external influences) before t_5, then there will be another forced withdrawal leading to a new plateau, etc.

Mechanical systems with equilibrium plateaus that are long compared to the times between interruptions are restricted to forced withdrawals and young parts of plateaus to the exclusion of spontaneous withdrawals and old parts of plateaus. The system is kept young so to speak, by being "kicked around" by frequent outside changes. Old equilibria and their spontaneous processes become excluded. The system becomes irreversible in the sense of criterion 3.

The mechanism described above could be the origin of entropic irreversibility for fluids. The test of the conjecture is whether or not the separation of young equilibria and forced withdrawals from old equilibria and spontaneous withdrawals agrees with the separation of observed processes, which create entropy, from their absent reverses, which destroy entropy.

The excluded processes are the time reverses of the allowed processes. We assume the time reversibility of all processes (criterion 2) and we can follow the variation in Fig.7.2b backwards to find spontaneous processes at the ends of plateaus as the time reverses of the forced withdrawals at the beginnings of plateaus. But how do we show that the allowed forced withdrawals create entropy? (If we could show that, it would follow that the excluded reverses destroy entropy.)

Here statistical mechanics can enter the discussion in a useful role. Up to this point the mechanism of irreversibility has been described in purely mechanical terms. Irreversibility has already been found in the exclusion of certain processes. It was not necessary, as claimed by Montroll (Note 25), to "play statistical games in order to extract irreversible characteristics from reversible dynamical equations." The question that remains is whether or not entropy is created in allowed processes, and for that question we can use statistical mechanics as a guide.

According to statistical mechanics, as argued in the previous section, entropy is created in processes which are sufficiently reproducible to be predictable by statistical methods. And those processes are the forced (young) withdrawals from young equilibria, which happen also to be the allowed processes. In other words, statistical mechanics necessarily deals with the most probable situation, the most probable distributions, the most probable ensemble, the most probable macroscopic behavior consistent with prescribed values of a few macroscopic properties, and so it must be restricted (in its applications) to young equilibria, with their assurance of continued randomness, and to early forced withdrawals from such equilibria. But that restriction is just what we need. It enables us to conclude that entropy is created in the allowed young forced withdrawals, and hence destroyed in their reverses, which are the old excluded, spontaneous withdrawals.

We have found a mechanism for entropic irreversibility. Any mechanical system with equilibrium plateaus that are long compared to the times between substantial outside interruptions will exhibit a separation between allowed and excluded processes. In cases where entropy is a recognized property, the allowed processes will create entropy while their excluded reverses would destroy entropy.

It is hard to believe that there could be such a simple answer to questions about the origin of entropic irreversibility. Merely a matter of time scales for internal change vs external change. But that is our argument and we have that in mind for all that follows.

7.2 The Role of Time Scales

Consider a system of few or many particles with or without interparticle interactions which is kept isolated from changes in external forces for an indefinite time. The system is confined to a fixed volume. Such a system (Newtonian) would exhibit a periodic pattern of behavior over and over, and there is no reason to believe that the pattern could not occur in reverse. The system may or may not have equilibrium plateaus in the time variation of observed properties.

If there are enough particles to have noticeable equilibrium plateaus in the time variation of the properties of interest, then there will be a distribution of plateau lengths, depending on initial conditions. Little is known about the distribution of plateau lengths in general, and there is a need for a science of plateaus. But there are reasons to believe that if the plateaus are quite long compared to other times of interest, then the probability of a short plateau is very small. Let us refer to t_e (time at equilibrium) as a typical length for shorter plateaus. In Fig.7.2a, this is $t_2 - t_1$.

Next we introduce a typical time t_r for the relaxation of the system back to equilibrium after being far enough removed to be noticeably away from equilibrium. In Fig.7.2a this is $t_1 - t_0$. If the plateaus have negligible length, as in the case of a pendulum, for example, then t_r is poorly named as a "relaxation time," but we can simply treat it as the time during which the system is allowed to do its thing without outside interference.

Finally we abandon the restriction to isolated systems and allow for "interruptions" due to substantial changes in external influences. We could try to identify the typical time interval for the change, but that will carry us too far into the forest at this point, if we are not already too far, and so we assume that the interruptions are more or less instantaneous, but we do recognize the typical time t_1 between interruptions.

And so we have three times: t_e (at equilibrium and undisturbed), t_r (to approach a new equilibrium or a turning point in periodic motion), and t_i (between interruptions). There are six ways to order these times

$$t_1 < t_2 < t_3$$

and more ways if we want to distinguish between extreme differences and moderate differences, and more ways if we want to introduce other typical times. We can try speculatively to identify models or everyday situations that conform to each ordering. Needless to say, ambiguities and bad analogies abound. But we think it is profitable to try for a few cases.

Case 1: $t_e < t_r < t_i$

The system is essentially isolated (very long t_i) with negligible plateaus (very short t_e) and the time t_r of interest is spent in a condition of continual change. This is the predicament of the simple models in classical physics (oscillators, pendula, revolvers, ...). The model is periodic and no patterns of motion are excluded if we allow for all possible initial conditions. The system is reversible.

Case 2: $t_r < t_e < t_i$

Think of a usually isolated collection of coupled springs, a big bed spring or a big molecule with oscillating bonds. We could follow the distribution of vibrational energy over the system after an input from outside and notice that the energy becomes fairly uniformly distributed in a time t_r and stays that way for a time t_e. But then, at the end of t_e, the energy "spontaneously," with no outside cause, accumulates in some place. Maybe the molecule dissociates. (This idea is one of those used to explain "unimolecular" chemical reactions that proceed without collisions.) Or maybe the energy once again spreads out for another period of "equilibrium" and then periodically returns to the same location. The system in its predicament does have equilibrium plateaus, but they are shorter than the times between external interruptions, so the system can reach the old processes at the ends of plateaus. There is no exclusion of old processes. The system is reversible in the sense of criterion 3. And its periodicity is manifested.

Case 3: $t_r < t_i < t_e$

Interactions among systems provide interruptions for any one system at a frequency greater than the frequency of spontaneous withdrawals from equilibrium but less than the frequency of relaxation. Thus the system can respond to an interruption and reach a new equilibrium but it cannot stay at equilibrium long enough to reach an old process at the end of a plateau. Here we have an exclusion of old processes. The system is irreversible in the sense of criterion 3. This is the predicament of fluids, e.g., systems covered by the second law of thermodynamics.

Case 4: $t_i < t_r < t_e$

Here the system not only stays young (exclusion of old processes) but cannot even reach the end of a young process. It never relaxes all the way to equilibrium before being interrupted again. This is the predicament of living organisms, which has been described eloquently by Schrödinger (1944). According to Schrödinger the essence of life is being able to "move out" and attract interruptions at a frequency larger than the relaxation frequency. For example: breathing 16 times a minute rather than waiting for diffusion. Old processes never have a chance. The system is highly irreversible.

We could go on to other examples, but we notice already a common denominator for exclusion irreversibility:

$$t_i < t_e \tag{7.1}$$

We see that irreversibility is not an intrinsic property of a system, but a property of a predicament. In any predicament with typical times t_i, there will be a division of systems into those that are reversible ($t_e < t_i$) and those that are irreversible ($t_e < t_i$). The t_e are intrinsic, but not the t_i. Thus all systems with whatever t_e are potentially reversible in some environment. For fluids familiar to us, the t_e are so large that it is difficult to imagine a predicament in which they would be reversible. That adds to the mystique of irreversibility.

7.3 Exclusion and Dissipation

It has been argued above that the kind of irreversibility which is associated with entropy, which is the subject of paradoxes, which is associated with time's arrow, which is mystical on its surface, and which has been the subject of debate for a century, is actually exclusion irreversibility. That argument may have seemed to be fair enough.

And here we argue that exclusion irreversibility is a consequence of a simple relation between an intrinsic time t_e and an environmental time t_i (7.1). The argument may be flimsy in spots but it seems to go in the right direction. But now when we sit back and put those arguments together, we cannot help but think we have been had. After all, irreversibility, with all of our feelings about it, our sense of its powerful thrust, our association of irreversibility with dissipation, friction, decay, ... Who can convince us that such a force comes from a piddling relation like (7.1)? To focus that question, we ask for the relation between exclusion irreversibility and dissipation.

Consider a fluid contained in a finite vessel and let it be hit (compressed) at some point at an initial time. The disturbance (blip in the graph of pressure) would propagate out through the fluid, be reflected at the container walls, and continue propagating (moving). If the fluid were viscous, with frictional resistance everywhere, the size of the disturbance would decrease in time; it would be "damped." That is "dissipation." If the reduced blip reached a wall, it would be reflected, but would continue reducing and fade away (decay) into oblivion. An energy expert would say the energy had become spread out never to be localized again. A less sophisticated observer would simply say that the disturbance (wave) just gradually disappeared. Dissipation. Fact of life. Everything gets lost eventually.

But if we watched this fluid for a very long time during its isolation, the disturbance would reappear and again undergo its insulting demise. This is simply a matter of executing the periodic behavior predicted by Newtonian mechanics if it is isolated in the sense of very long t_i. Dissipation does not necessarily lead to irreversibility, nor to an absence of periodicity.

Suppose the fluid had no viscosity or any kind of frictional resistance to deformation. Maybe the blip could simply travel around being reflected from walls and never decay or die. No dissipation. Would the process be reversible?

In this case we could say that t_e is negligible. If we are watching the whole fluid, something will always be happening somewhere. In fact, the process is similar to the process of a single particle bouncing around from reflecting walls without energy losses, and it is certainly reversible. So long as we have the reverse of (7.1),

$$t_e < t_i \tag{7.2}$$

the process is reversible, whether t_e is large or small.

We have noticed that the introduction of a particular kind of dissipation into a model can drastically increase the time t_e. For a given time t_i, we could have t_e increase from smaller than t_i to larger than t_i, and have reversibility replaced by irreversibility. Perhaps our psychological tendency to associate irreversibility with dissipation is due, at least partly, to an intuitive awareness of the effect of dissipation on t_e.

But we must be on guard, for this kind of intuition can lead us astray. For example, a thin gas allowed to expand without collisions in a finite container would reach uniformity throughout the container in an instant while the same gas allowed to expand in a container filled with another gas which is dense would take forever to reach uniformity. A smoke ring puffed into the atmosphere within a dining room would linger many seconds and take hours to reach the other end of the room, but a smoke ring introduced into an evacuated vessel would disperse faster than the eye could follow. The "dissipation" is more severe in the fast collisionless case than in the slow diffusive case. Or is it? There we see some confusion about the concept of dissipation. We may argue that the concept becomes clearer if we move away from diffusion to viscosity. Viscous flow is dissipative, and the more viscous, the more dissipative. Actually, the viscosity of a thin collisionless gas is much greater than the viscosity of a dense gas. If we think that dissipation is enhanced by molecular collisions, we have it backwards. And if we think that diffusion is caused by collisions, a common misconception, we have it backwards. Diffusion is inhibited by collisions. It is very easy to become confused by wrong associations with the terminology of fluid dynamics. "Conductive" transport means "inhibited" transport. Inhibitive "dissipation" can mean "freedom to fly away," but it is also used to refer to inhibitive friction. To some, dissipation is associated with the loss of "structure." To others, it is the frictional resistance against that loss of structure.

Fortunately there is a systematic way to bring dissipation and irreversibility together, and this approach offers a clue to identity of the general circumstances under which the two can be related. It also offers to us the last clue we need to understand the origin of irreversibility in fluid behavior. We refer here to the Thermodynamics of Irreversible Processes.

According to irreversible thermodynamics, as we will see in the next chapter, the generalized idea of entropic irreversibility can be represented by a formula for the rate of irreversible creation of entropy (density):

$$\sigma = -\frac{1}{T^2}\sum_i q_i \frac{\partial}{\partial x_i} T - \frac{1}{T}\sum_{ij} (p_{ij} - p\delta_{ij}) \frac{\partial}{\partial x_j} v_i \qquad (7.3)$$

where $\underline{q} = (q_1, q_2, q_3)$ is the vector for the conductive flux of heat, and p_{ij} is an element of the pressure tensor and δ_{ij} is 0 if $i \neq j$ and 1 if $i = j$. The pressure p is what the pressure would be at equilibrium, and the velocity $\underline{v} = (v_1, v_2, v_3)$ is the fluid velocity. This rate contains no contribution from the accumulation of entropy to the mere convection of entropy by moving fluid carrying its entropy with it. It does not even contain the contribution (reversible) due to heat conduction. It contains only the contribution to entropy creation due to processes that are classified as irreversible and also classified as dissipative.

In irreversible thermodynamics it is important to be able to see a certain pattern in the right side of (7.3)

$$\sigma = \sum_i J_i X_i \qquad (7.4)$$

where the J_i are "fluxes" and the X_i are "forces." The forces could be negative derivatives of T and gradients of components of $\underline{v}$, which "drive" the conduction of energy and momentum within the fluid. The fluxes could be primarily components of the heat flux $\underline{q}$ or elements of the momentum flux (pressure) tensor, with something subtracted off (eliminating the nondissipative effects of expansion and contraction). Without going into the details here we point to (7.4) as a simple, compact formula that somehow contains the systematic summary of the forces and fluxes associated with irreversibility and dissipation. There is such a thing, as we will see in more detail later.

The point to be made here is that the development of irreversible thermodynamics up to and including the formulas

(7.3) and (7.4) proceeds without introducing dissipation or irreversibility. That introduction comes in the next step, and we especially want to see what that next step is. As for (7.3) and (7.4), they are not irreversible in any sense by anyone's criteria. And there is no hint of dissipation. The so-called entropy "creation" (7.3) can be positive or negative. The formula describes equally well the creation of entropy in young forced withdrawals from equilibrium and the destruction of entropy in old spontaneous withdrawals and the creation or destruction of entropy in any other process we can imagine for the fluid or the molecular aggregate.

Then we have the "next step" in the logic line of irreversible thermodynamics which suddenly introduces irreversibility and renders (7.4) as dissipative. In this step we relate the fluxes J_i to the forces X_i by means of Fick's law for diffusion. Newton's law for momentum transport (viscosity), Fourier's law for heat conductivity, Ohm's law for electrical charge conductivity, etc. In each case, the flux J_i is proportional to the force X_i. More generally, we have each flux as a linear combination,

$$J_i = \sum_i L_{ij} X_j \tag{7.5}$$

as described in Chapter 4, and explained in the next chapter.

It is (7.5) that brings in all the irreversibility and dissipation. It is (7.5) with its development that proves entropy creation to be positive (irreversibility) and backs up the best intuition about dissipation. So it is to (7.5) that we must go to find the origin of associations between dissipation and irreversibility.

What we will find in later chapters is that (7.5) holds only under highly restricted conditions, and those must be the conditions that bring together dissipation and irreversibility. Those conditions are, as we will see, easy to identify. The system must be leaving a recent equilibrium. The conditions hold for the early stage of forced withdrawal from a short-lived equilibrium. This is especially the predicament of fluids.

7.4 The Principle of Recent Equilibrium

Focusing now on the predicament of actual fluids, we see that they have such long t_e as to be irreversible with or without dissipation. They are "strung out," so to speak, and in a condition totally at the beck and call of dictatorial interruptions at t_1, with overwhelming irreversibility in the sense of (7.1). What makes them dissipative is that they are always leaving equilibrium, never approaching (except by default as administered by the next interruption). They are always "young" with increasing entropy. It is this condition of recent equilibrium that brings together all of our formulas and prejudices about fluid behavior and correlates irreversibility with dissipation.

Any time we encounter a fluid it is either at equilibrium, and a short-lived equilibrium, so we are guaranteed a long continuation of equilibrium, or else it is in a young forced withdrawal from a recent equilibrium. And that recent equilibrium was so short-lived, in turn, that we can assume it would have continued for a long time if not interrupted by the forced withdrawal. Such a fluid is always near in time to a recent equilibrium. It may or may not be near "in space" (i.e., small spatial gradients) to any equilibrium, but it is always near in time to a past equilibrium.

The common preoccupation with the "approach" to a future equilibrium should be replaced by a focus on the past equilibrium. All of the intrinsic properties of the system evolve from that. One can watch diffusion with Fick's law through a nearly infinite gradient and with great accuracy so long as that infinite gradient was produced in the recent past by a change of external forces acting on a system that was at equilibrium against the earlier forces. And the same with other so-called irreversible processes.

We can summarize these arguments in a statement of the Principle of Recent Equilibrium:

> All fluids encountered in nature are at a young equilibrium or else are in a forced withdrawal from a young equilibrium. This is the origin of their irreversibility and the cause of their dissipation.

7.5 A Reflection

The argument above to the effect that dissipation is due to being forced to leave equilibrium is still shy, we must acknowledge. Maybe here we should try for some new metaphors. Rather than watching a system being forced to leave equilibrium we should see it as "released" from equilibrium and being suddenly allowed to exhibit its bee-hive of molecular phenomena, its infinite variety of potential evolutions, all because of the loss of old constraints.

But the old constraints are replaced by new constraints. (There are always constraints among interacting systems.) Thus the result of the loosening of old constraints, which led to the unreversible "flight to freedom," was soon, though gradually, brought to a halt by the grinding dissipation of randomization against the new constraints. In this sense we can see the point of focusing on the approach to a new equilibrium (determined by the new constraints). Our previous denials of the importance of future equilibria may have been overstated.

Nevertheless, it is the previous equilibrium that determines rates of change, forms of macroscopic equations, including the "reciprocal relations" (4.3), and the positiveness of entropy creation. So we do stick to our arguments about the origins of irreversibility and dissipation but acknowledge that the maintenance of the continual nature of these characteristics is dependent on the continuous arrival of new constraints.

CHAPTER 8

IRREVERSIBILITY IN FLUID DYNAMICS

Let us begin with a broad view of fluid behavior and then isolate the part which is dependent on the cause of irreversibility. By "fluid" we mean any macroscopic and inanimate sample of matter, solid, liquid, or gas. It is true that solids are not very "fluid," but they do flow slowly and they do the other things (melt, conduct, react ....) covered by theories about "fluid dynamics," so we have no reason to exclude them here. Nor do we exclude plasma (ionized gases), often referred to as a fourth state of matter, or surface films or even rarefied gases. But we do restrict ourselves at every opportunity to the simplest fluid predicament that can offer the needed examples. We are not trying to construct a fluid dynamics but merely to identify the cause of the irreversible part.

By "fluid dynamics" we mean a set of equations which is sufficient to predict the pattern of behavior under discussion, or, more generally, the big set of equations from which such subsets can be drawn. In any case, to have a dynamics, i.e., a predictive recipe, it is necessary to have as many equations to be solved as unknowns to be solved for, and to have the necessary auxiliary data including boundary and initial conditions.

8.1 The Fluid Concept

By "fluid" we mean a continuum with properties, such as the mass density, which vary too slowly in space to be interpreted easily as reflections of molecular properties. The common idea of mass density as a limiting value ρ of the ratio of mass to volume for a shrinking sample,

$$\rho = \lim_{V \to 0} \left(\frac{M}{V}\right)$$

is not relevant. Such a density would jump from zero to very large values as the point where it is evaluated moves from outside an atom to inside an atom, or should we say, nucleus. It would be better to take the succession of smaller samples with decreasing volumes V_1, V_2, ... and masses M_1, M_2, ..., all containing the point at which we intend to evaluate the density, and then to plot the ratios as in Fig.8.1. For samples too small the ratio M/V is

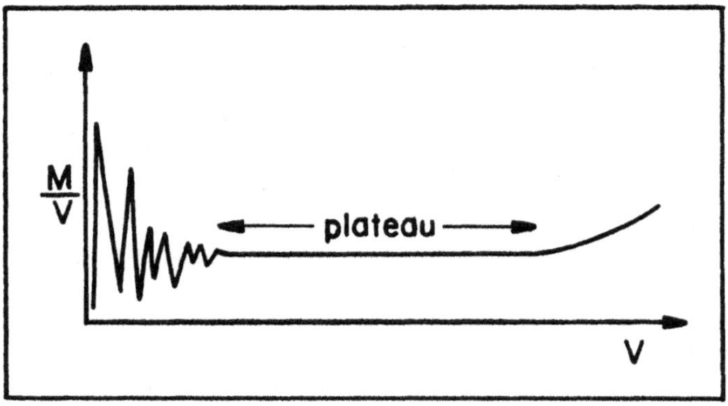

Fig.8.1 The Plateau of Macroscopicity

erratic. For samples too large the ratio varies with the sample chosen. Hopefully there is a long plateau extending through the volume range of measuring devices so that we can take the plateau value as the unique and reproducible value of fluid density at the position of interest. We know from experience that such plateaus are common. It is these plateaus which allow us to develop, without contradictions, the relationship between macroscopic and the molecular concepts of matter.

The relationship between fluids and molecules becomes more complicated when we consider the fact that a continuous fluid has motion correlations that do not exist for molecules. Pieces of fluid very near a given piece move nearly parallel to that given piece while the underlying molecules are buzzing around in all directions. The velocities of nearby molecules are not correlated. The correlation which emerges at the fluid level is most remarkable. We can follow a piece of fluid through its motions, elongations, whirls, and stagnancies, assuming all the while that the piece retains its "integrity," remaining always the same piece, without contamination, and yet the molecules are leaving the game with new substitutes coming in all the time. A piece of fluid can accelerate merely because the substitutes are faster.

The continuous nature of fluids can be explained partly by the collisional inhibition that keeps molecules close together for awhile. The molecules in a cloud or a smoke ring, for example, are moving at thousands of miles per hour, but they do not get far fast because of the hundreds of thousands of collisions per second. Another factor contributing to the continuous nature of fluids is the correlation of accelerations due to external forces. These correlations are the same at the molecular and fluid levels. For example, the convective (collective) rising of hot air in gravity, which is said to be due to "buoyancy" at the macroscopic level, is due to the larger "fall" of a cold (slow) molecule compared to the smaller fall of a hot (fast) molecule in the same travel distance. (The fall due to gravity is the same for both molecules in the same time, but not in the same travel distance between collisions.) A third factor contributing to the continuity of fluids is the plateau nature of the measurable properties. We can say all this. But in the end we still have to marvel at the possibility of fluidity. It must have a strong dependence on the large numbers of molecules involved and the tremendous differences in scales for time and distance.

One way to get to the edge of the fluid concept is to observe the behavior of rarefied gases. The collisional inhibition to wild molecular motion is more or less missing. This led a generation of workers in rarefied gas dynamics to refer to "free molecule flow" as distinct from fluid flow. But they sought and found fluid equations. Ironically, some fluid equations, such as Fick's law for diffusion, can be derived more exactly for rarefied gases than for dense gases (Biolsi, 1974). In any case, even rarefied gases have a fluid nature.

The evolution of fluid behavior can be described by mathematical equations. Although we are not especially interested in the mathematics of fluid descriptions, we do find clues in the equations and their origins which help us to understand the irreversible nature of fluid behavior. We will examine some of the equations below. We will find that although the fluid concept itself does not require the causes of irreversibility, nevertheless those causes are extremely important in determining the dynamics of observed fluid behavior. First we classify patterns of fluid behavior in a convenient way (not exhaustive) and then we classify the equations, hoping to isolate the features of the equations that represent the causes of irreversibility.

8.2 Fluid Processes

The various patterns of behavior which have been found for fluids are often classified as follows:

a) Equilibrium
b) Steady State
c) Change

A fluid at equilibrium is "stagnant" and constant in time in every way. Not only are the property values constant, but there is no motion, no motion of fluid, no motion of heat, no motion of anything. The fluxes are zero. If there are no changes in external influences, then there will be no change of any kind in the macroscopic condition of the fluid. Whatever the molecules are doing, it all cancels out at the macroscopic level. Maybe not quite all. The temperature of the fluid reflects the motions of the molecules, but it too remains constant. If there are phases at chemical equilibrium, then molecules can be moving from one phase to another or changing to form new molecules, but opposite processes at the molecular level must have reached the same rates, for there is no change at the macroscopic level.

There are other processes in which no property changes in time even though there is a macroscopic motion such as the conduction of heat from one place to another. Or the conduction of charge or a chemical species or momentum. Conductive fluxes can have nonzero values which remain constant in time. The fluid itself can be flowing with constant velocity and no other change. Such patterns of behavior are referred to as "steady states."

Finally there are processes in which there is not only a flux (or fluxes) due to conduction or convection but also temporal changes in the values of densities and fluxes. Then we have the full-fledged time-dependent change. Often such changes are short-lived and are called "transients" or "relaxations" or "responses," etc. But they can also be long-lived patterns of change.

8.3 Fluid Equations

The classification of fluid processes described above can be extended with some modifications to the equations used to describe the processes. We will use the following classification:

a) Equations of Equilibrium
b) Equations of Flux
c) Equations of Change.

We will separate equations of flux into two groups, those that are "simple" and those that are "complex." The reason for the separation will be obvious. We will also separate the equations of change into those that are "fundamental" and those that are "stochastic." The reason for that separation will turn out to be the heart of the whole discussion here about fluid dynamics. The stochastic equations, including all of the equations of flux and some of the equations of change, are the equations that are irreversible. The origin of their irreversibility lies in the equations of equilibrium. Only the fundamental equations of change escape from exclusion irreversibility.

For any particular pattern of change, we are ready to accept any set of equations of these types which is mathematically "complete" and physically "deterministic" in the sense of being fully predictive or "dynamical." We will accept that and call it a "dynamics" and then try to say where irreversibility comes into the dynamical recipe. But before we can get to questions about irreversibility we must have a dynamics. Any set of equations which has more unknowns than equations is not dynamical, not deterministic, and we cannot easily assess the presence of irreversibility in such a set.

From the accepted pedogogical point of view it is desirable to begin the study of fluid dynamics with equations of equilibrium and then go on to fluxes and eventually to temporal change. There are good reasons for this order. Equations of equilibrium usually involve no derivatives; e.g., $pV = nRT$. They can be used without calculus. Flux equations, such as Fick's law of diffusion, involve spatial derivatives or "gradients," which can be approximated without calculus. Equations of change come after calculus with considerations of boundary and initial conditions. In this order there is an increasing demand on mathematical confidence. Furthermore, there are "reductions" which go the other way. The full-fledged set of equations for fluid dynamics can be reduced to a set of equations for steady states by eliminating time derivatives, and the set of steady-state equations can be reduced to a set of equilibrium by eliminating fluxes. So it is desirable in some ways to consider the equations of fluid dynamics in the order presented above.

But this sequence reverses the order of our needs. There is no irreversibility in the fundamental equations of change. Irreversibility creeps in when we study stochastic equations of change and equations of flux. Then irreversibility exhibits its full nature at equilibrium. And so we are led to examine the equations in an order opposite to the order given above.

8.4 Fundamental Equations of Change

These equations are derivable directly from Newtonian equations governing the motions of particles or from thermodynamic equations applied to uniform bodies. By "body" or "particle" we mean a uniform, rigid thing, a nondeformable piece of matter which has no conductions or other noticeable internal change. By thermodynamics we mean the first and second laws without that part of the second law which says entropy must increase. The second law in its purest form merely says there is such a thing as entropy, i.e., that reversible dq divided by temperature is an exact differential and hence $dS = dq_{rev}/T$ is the variation of a state function like volume or energy.

Suppose we have a pure fluid with properties that vary only in the x-direction,

$$\begin{aligned}
\rho = \rho(x,t) &= \text{mass density} \\
v = v(x,t) &= \text{fluid velocity} \\
p = p(x,t) &= \text{fluid pressure} \\
q = q(x,t) &= \text{conductive heat flux} \\
u = u(x,t) &= \text{density of internal energy} \\
s = s(x,t) &= \text{density of entropy}
\end{aligned} \tag{8.1}$$

The fundamental equations of change expressing, for a piece of fluid, the conservation of mass is

$$\frac{\partial}{\partial t}\rho + \frac{\partial}{\partial x}\rho v = 0 \tag{8.2}$$

Newton's F = ma is

$$\frac{\partial}{\partial t}\rho v + \frac{\partial}{\partial x}\rho vv = -\frac{\partial}{\partial x}p + \rho F_e \tag{8.3}$$

The first law of thermodynamics is

$$\frac{\partial}{\partial t}u + \frac{\partial}{\partial x}uv = -\frac{\partial}{\partial x}q - p\frac{\partial}{\partial x}v \tag{8.4}$$

and the second law of thermodynamics:

$$\frac{\partial}{\partial t}s + \frac{\partial}{\partial x}sv = -\frac{1}{T}\frac{\partial}{\partial x}q - \frac{1}{T}(p - p_e)\frac{\partial}{\partial x}v \tag{8.5}$$

The quantity F_e is the external force per unit mass, which is ignored in Eqs.(8.4) and (8.5). The pressure p_e in (8.5) is what p would be if the fluid were in equilibrium with the same ρ and u.

These equations have certain general features which deserve our consideration, but before we go to that let us review the derivation of the equations. They can be derived directly from molecular equations without any specific knowledge of the molecular arrangements. In other words, they follow ultimately from the basic laws of molecular dynamics, as we will see in the next chapter. They can also be derived even more directly by applying the laws involved to a piece of fluid regarded as a particle. The latter derivation is worth our consideration in more detail.

Suppose we have a pure fluid with mass density ρ(x) which varies in the x-direction. In fact, for the moment, let ρ be the mass per unit length in the x-direction. Then ρdx is the mass of a sample of length dx along the x-axis. To be more precise we consider the profile of ρ(x,t) at time t, as in Fig.8.2.and say that the mass of the sample between x and x+dx is the shaded area under the curve, which can be approximated for the purposes of calculus by

$$\text{sample mass at time } t = \rho(x,t)\,dx\,. \tag{8.6}$$

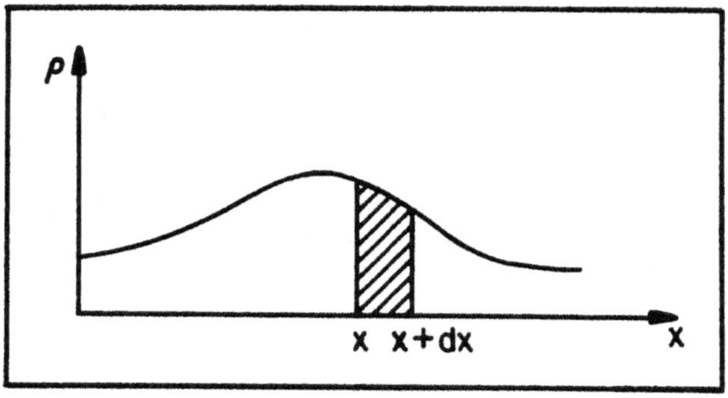

Fig.8.2 A Piece of Fluid

As time passes the profile in Fig.8.2 changes and so do the positions x and x + dx of the extremities of our sample. But the mass of the fluid between these changing extremities remains constant, for we assume that our piece of fluid retains its identity no matter where it goes or how much it stretches or contracts. It may actually trade molecules with the surrounding fluid, but at the fluid level, speaking conceptually, it remains the same continuous blob it always was. It has "fluid integrity."

In order to describe precisely what happens to our blob of fluid we have to introduce an intermediate "helping function" which is to be eliminated eventually from the description but which provides the conceptual vehicle for derivations. The need for this function can be illustrated immediately by noticing the difference between the velocity of a particle at time t and the velocity of a "fluid particle" (point in the fluid) $v(x,t)$ at position x and time t. The former velocity can be regarded as the derivative dx/dt of $x(t)$. But the fluid velocity can be no such thing, for x and t are independent variables. It is true that we know what is meant by fluid velocity, thinking of it, perhaps, as the velocity of a small drop of dye in the fluid, the motion of which can be followed visually before the drop diffuses too much. But how do we express that idea in mathematical terms? We need a "helping function."

One way to define a suitable helping function is as follows. Suppose the fluid particle, or drop of dye, originally at x at time t, moves in a subsequent time τ to the new position x'. Thus $x'(x,t,\tau)$ is a function of the initial position x, the initial time t, and the elapsed time τ. We may never have a formula for this function, but it will serve as an aid in derivations. For example, we can express our concept of fluid velocity $v(x,t)$ as

$$v(x,t) = \left[\frac{\partial}{\partial\tau} x'(x,t+\tau)\right]_{\tau\to 0} \tag{8.7}$$

Whether or not that formula turns out to be useful (it does), it says exactly what we mean by "fluid velocity." We will refer to the formula $x'(x,t+\tau)$ as the "fluid trajectory."

Now going back to our blob of fluid of length dx between x and x+dx, we may say that the extremity x evolves in time τ to $x'(x,t+\tau)$ and the other extremity $x+dx$ evolves to $x'(x+dx,\ t+\tau)$. Thus the length dx evolves to

$$dx' = x'(x+dx,\ t+\tau) - x'(x,t+\tau) \tag{8.8}$$

According to (8.7), the rate of change of this length is

$$\frac{\partial}{\partial\tau} dx' = v(x+dx,\ t+\tau) - v(x,t+\tau) \tag{8.9}$$

Backing up to the time t of interest, $\tau\to 0$, we find the current (instantaneous) rate of change of the length dx as

$$\frac{d}{dt} dx = v(x+dx,t) - v(x,t) = dx\,\frac{\partial v}{\partial x} \tag{8.10}$$

where the symbol d/dt is to be interpreted as the rate of change as seen by an observer moving with the fluid element, i.e., the rate of change for our permanent fluid blob followed in its evolution. The manipulation toward the rightmost side of (8.10) is based on the definition of a derivative,

$$\frac{\partial y}{\partial x} = \frac{y(x+dx) - y(x)}{dx} \tag{8.11}$$

where here y is v. We notice that the helping function x' explicit in (8.9) has disappeared in (8.10) and we are left with the x-derivative of $v(x,t)$, which is meaningful without the helping function. In fact, one could argue that the rate of change of dx is obviously the difference between the velocities of the extremities at x and x+dx. We did not need the argument involving the helping function.

If such a strong intuition continues to prevail, we will never need the helping function. But maybe we should admit that the x-derivative of v(x,t) being the rate of expansion (or contraction) of our fluid blob was not so obvious.

This rate of expansion or contraction in three dimensions is called the "divergence" of the fluid at the point $\underline{r} = (x, y, z)$ at time t. For a fluid element of volume

$$d\underline{r} = dx\,dy\,dz \tag{8.12}$$

we have

$$\frac{d}{dt}\,d\underline{r} = d\underline{r}\left(\frac{\partial}{\partial \underline{r}} \cdot \underline{v}\right) = d\underline{r}(\underline{\nabla} \cdot \underline{v}) \tag{8.13}$$

In this expression we have used the notational device of (8.12) and the other devices listed below.

<u>Notational devices</u>: If $\underline{a} = (a_1, a_2, a_3)$ or (a_x, a_y, a_z) is a vector, we underline it and call it $\underline{a}$. And if two such vectors are multiplied to form the "scalar product" $\underline{a} \cdot \underline{b}$, i.e., the "dot" product, which is the magnitude of $\underline{a}$ times the projection of $\underline{b}$ in the direction of $\underline{a}$, then we have

$$\underline{a} \cdot \underline{b} = a_1 b_1 + a_2 b_2 + a_3 b_3 \tag{8.14}$$

in which the numerical subscripts could be replaced by x, y, and z. Thus the vector $\underline{v}$ is

$$\underline{v} = (v_x, v_y, v_z)$$

and the "gradient" vector is

$$\frac{\partial}{\partial \underline{r}} = \left(\frac{\partial}{\partial x}, \frac{\partial}{\partial y}, \frac{\partial}{\partial z}\right) = \underline{\nabla} \tag{8.15}$$

Thus the "divergence in (8.13) above, without the $d\underline{r}$, is

$$\underline{\nabla} \cdot \underline{v} = \frac{\partial}{\partial x} v_x + \frac{\partial}{\partial y} v_y + \frac{\partial}{\partial z} v_z \tag{8.16}$$

We need one more idea based on the trajectory x' of a hypothetical fluid particle, and then we can proceed to the derivation of the fundamental equations of change without further interruptions. Suppose we have a fluid property f(x,t) defined at each point (x,t). What is its rate of change as seen by an observer moving with the fluid, or as followed as the rate of change for our evolving fluid element? Using our helping function and the chain-rule of differentiation, we find

$$\frac{d}{dt} f(x,t) = \left[\frac{\partial}{\partial \tau} f(x',t+\tau)\right]_{\tau=0} = \left[\frac{\partial f}{\partial x'}\frac{\partial x'}{\partial \tau} + \frac{\partial f}{\partial (t+\tau)}\right]_{\tau=0}$$

$$= \frac{\partial f}{\partial t} + v\frac{\partial}{\partial x} f \qquad (8.17)$$

Equations (8.10) and (8.17) give us the "kinematics" of fluid dynamics, i.e., the way to express the rate of change of properties of interest. These properties are the extensive properties, such as the mass, energy momentum, internal energy, enthalpy, etc. Any such property can be evaluated as a density of the property f multiplied by the length dx of the fluid element. (In two and three dimensions, the length dx is replaced by the area dx dy or the volume dx dy dz.) The rate of change of the property for the evolving element is

$$\frac{d}{dt} f(dx) = \left(\frac{\partial f}{\partial t} + v\frac{\partial f}{\partial x} + f\frac{\partial v}{\partial x}\right) dx = \left[\frac{\partial f}{\partial t} + \frac{\partial}{\partial x}(vf)\right] dx \qquad (8.18)$$

The sum in the right-most brackets, i.e., the partial time derivative plus the spatial derivative of vf, will become for us a familiar pattern for the left-hand sides of equations of change.

If the quantity of interest is mass, so that the particle equation is

$$\frac{d}{dt} m = \frac{d}{dt}(\rho dx) = 0 \qquad (8.19)$$

then the corresponding fluid equation is (8.19) with f as the mass density ρ and the rate of change set equal to zero, even if dx is not quite zero,

$$\frac{\partial}{\partial t}\rho + \frac{\partial}{\partial x}(v\rho) = 0 \qquad (8.20)$$

This is Eq.(8.2) which is often called the "equation of continuity." In three dimensions the equation of continuity is

$$\frac{\partial}{\partial t} \rho + \underline{\nabla} \cdot (\underline{v}\rho) = 0 \tag{8.21}$$

If the quantity of interest is the momentum, so that the particle equation is the Newtonian equation

$$ma = \frac{d}{dt} (mv) = F \tag{8.22}$$

then f is the momentum density ρv and we have the fluid equation as

$$dx\left[\frac{\partial}{\partial t} (\rho v) + \frac{\partial}{\partial x} (\rho vv)\right] = F \tag{8.23}$$

This formula makes explicit only the left side of (8.22). We must now express the right side, i.e., the force, in terms of fluid properties. The task is accomplished in theories about fluid dynamics by separating the force F on the fluid element into an "external" part and an "internal" part,

$$F = F_{ext} + F_{int} \tag{8.24}$$

The external part is the force external to the fluid acting at a distance on our fluid element. The formula can be assumed to be known from the description of such a force acting on discrete particles. Here we may say, for convenience, that F_{ext} is that force per unit mass multiplied by the mass ρdx of our element,

$$F_{ext} = F_e \rho dx \tag{8.25}$$

The internal force on our element due to the pushing of the fluid fore and aft to the element can be described in terms of a force magnitude $p(x,t)$, which is the magnitude of the fluid force in either direction at a point x at time t. The forces in both directions at x must have the same magnitude or the fluid would be split up by the intrusion of "vacuum bubbles," and that goes against the integrity of the fluid concept. The force to the right at x pushing our element minus the force to the left at $x + dx$ is the net internal force on our fluid element:

$$F_{int} = p(x,t) - p(x+dx,t) = - dx \frac{\partial}{\partial x} p(x,t) \qquad (8.26)$$

Here we have used, again, the idea of (8.10). If we put together what we have in Eqs.(8.23) - (8.26), and then cancel out dx, we find

$$\frac{\partial}{\partial t} (\rho v) + \frac{\partial}{\partial x} (\rho v v) = - \frac{\partial}{\partial x} p + F_e \rho \qquad (8.27)$$

This is the same as (8.3) and is called the "equation of motion" in fluid dynamics.

The internal fluid force (8.26) has no exact counterpart in particle dynamics. A discrete particle does not experience "touching" forces. But the fluid force can be understood in terms of the underlying molecular model. It includes, for example, the action-at-a-distance forces exerted on molecules within the fluid element by molecules outside the element. And it includes the accelerating (or decelerating) effect of sending in faster molecules to replace leaving slower molecules (or vice versa) in our fluid element. Molecules do flow into and out from our fluid element and alter the momentum of the element. (Here again we encounter the miracle of fluidity and have to ponder the compatibility of fluid and molecular models.)

The three-dimensional version of (8.27) is three equations, one for each direction, which can be expressed collectively, in vector notation, as

$$\frac{\partial}{\partial t} (\rho \underline{v}) + \frac{\partial}{\partial \underline{v}} \cdot (\rho \underline{v}\,\underline{v}) = - \frac{\partial}{\partial \underline{r}} \cdot \underline{\underline{p}} + \underline{F}_e \rho \qquad (8.28)$$

The second term on the left side can be understood in terms of the vector notation expanded as follows:

$$\frac{\partial}{\partial \underline{r}} \cdot \rho \underline{v}\,\underline{v} = \left(\frac{\partial}{\partial \underline{r}} \cdot \rho \underline{v}\right)\underline{v} = \left(\frac{\partial}{\partial x} \rho v_x + \frac{\partial}{\partial y} \rho v_y + \frac{\partial}{\partial z} \rho v_z\right)\underline{v} \qquad (8.29)$$

Here the differentiations apply to every function to the right. The first term on the right side of (8.28) requires a lot of explanation, both notational and conceptual. It is desirable in this quick search for signs of irreversibility within fluid dynamics to pass by such details. Furthermore, a great deal of fluid dynamics is developed mathematically without confronting that first term on the right side of (8.28). If the fluid viscosities can be ignored, then that term is merely the gradient of a scalar (and familiar) pressure p,

$$\frac{\partial}{\partial \underline{r}} p = \underline{\nabla} p \tag{8.30}$$

But here we have to notice viscosity and its "dissipating" influence. It is widely regarded as one of the main "causes" of irreversibility, like "friction" in particle or body dynamics.

Consider the internal fluid forces acting to accelerate our fluid element to the right along the y-axis, as indicated in Fig.8.3. Before we identify the force on the element we

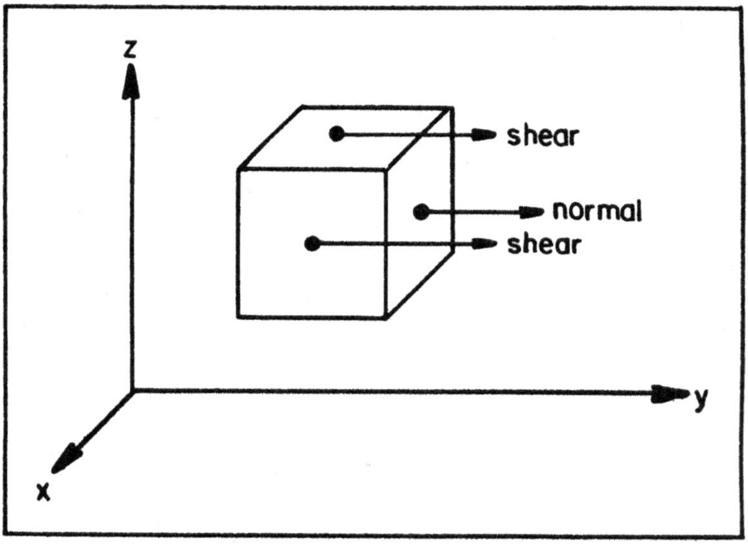

Fig.8.3 Forces on a Fluid Element

need to identify the forces acting at a point (x,y,z) in a fluid at some particular time. These forces are so complicated that we have to divide them into three groups. Each group operates on a plane (infinitesimal) centered at the point of interest. Let us consider a "z-plane" perpendicular to the z-axis with area dx dy, as in Fig.8.4. The force normal (perpendicular) to that plane, pushing upwards, call it F_{zz}, where the second z gives the direction of the

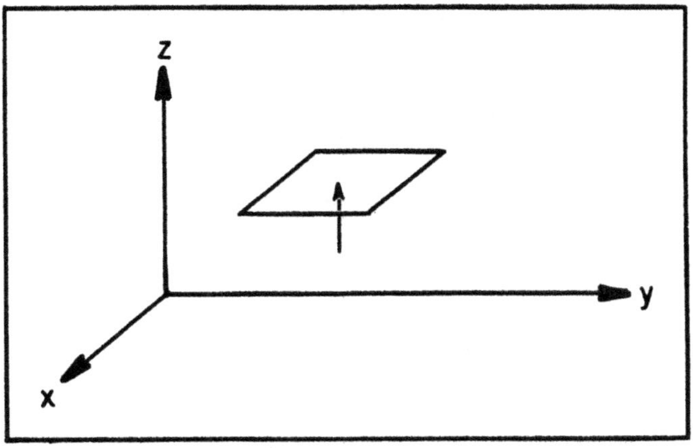

Fig.8.4 Normal Pressures on a z-Plane

force and the first z identifies the z-plane, as opposed to an x-plane or a y-plane; that force F_{zz} can be expressed as a pressure p_{zz} multiplied by the area dxdy.

There is also a "shearing" force trying to move the planar element to the right in the y-direction, as in Fig.8.5, which can be formulated as p_{zy}dxdy. By convention, we use the positive sign for the shearing force exerted by

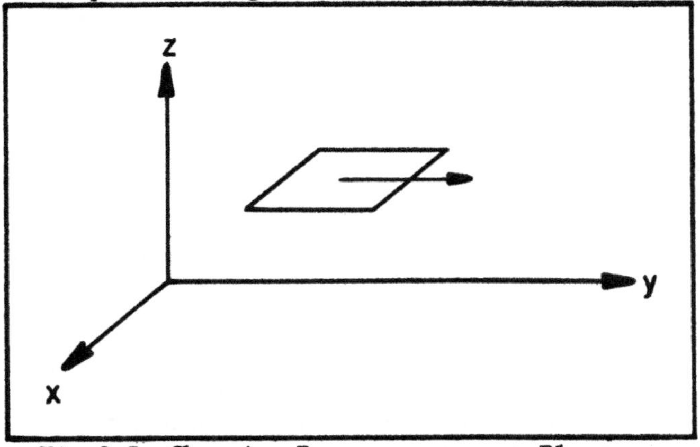

Fig.8.5 Shearing Pressures on a z-Plane

the fluid <u>below</u> the area element. (If that force actually goes the other way, then p_{zy} will be a negative number.) It turns out to be convenient to take p_{zy} positive in the negative y-direction when referring to "shear from above," which implies that p_{zy} is a negative number if the actual shear from above is in the positive y-direction (Note 26).

What has been said for shear in the y-direction can be repeated for shear in the x-direction. And so we have p_{zx} as the "shearing pressure" on a z-plane in the x-direction. In the mechanics of hard-to-deform bodies these shearing pressures are referred to as "stresses." In any language, the complete set of these stresses or pressures is a collection of nine numbers which we will call the "pressure tensor:"

$$P = \begin{pmatrix} p_{xx} & p_{xy} & p_{xz} \\ p_{yx} & p_{yy} & p_{yz} \\ p_{zx} & p_{zy} & p_{zz} \end{pmatrix} \tag{8.31}$$

The force in the y-direction on our fluid element is due to a difference of normal pressures on the y faces, fore and aft,

$$\left[p_{yy}(x,y,z) - p_{yy}(x,y+dy),z)\right] dx\,dz = -\frac{\partial}{\partial y} p_{yy}\, dxdydz \tag{8.32}$$

and a difference of shearing stresses on the z-faces and x-faces, similarly,

$$\left(-\frac{\partial}{\partial z} p_{zy} - \frac{\partial}{\partial x} p_{xy}\right) dx\,dy\,dz \tag{8.33}$$

so we have, altogether,

$$\text{force in y-direction} = -\left(\frac{\partial}{\partial x} p_{xy} + \frac{\partial}{\partial y} p_{yy} + \frac{\partial}{\partial z} p_{zy}\right) dx\,dy\,dz \tag{8.34}$$

The formula (8.34) gives the y-component of the first term on the right side of (8.28). The dot notation for the multiplication in (8.28) is explained by the notation

$$(\underline{a} \cdot \underline{\underline{b}})_i = \sum_j a_j b_{ji} \tag{8.35}$$

So far we have considered the conservation of mass and the Newtonian relation between forces and momentum change as as applied to a fluid element and we have the corresponding fluid equations of change. In one dimension these equations are (8.20) and (8.27). We notice that (8.20) by itself is not "dynamical," not deterministic, not solvable. It is one equation for two unknowns, $\rho(x,t)$ and $v(x,t)$. The evolution of these properties from initial conditions is not determined. All evolutions obey the equation but none is selected by the equation. When we add Eq.(8.27), hoping perhaps to find a dynamical system of two equations for two unknowns, we are disappointed, for (8.27) brings with it a third unknown, the pressure $p(x,t)$. In three dimensions the situation is even worse. We have (8.21), which is one equation for four unknowns, ρ and the three components of $\underline{v}$, and we have (8.28), which is a set of three equations that adds nine unknowns to our growing list of unknowns, namely, the nine components of the pressure tensor (8.31). If there is no viscosity we can reduce the number of unknowns in three dimensions to five, ρ, $\underline{v}$ and the scalar pressure p, but we still only have four equations. We could generate more equations by replacing our pure fluid by a mixture of fluids. Then we could have several equations of continuity and new equations of motion, etc. But the number of unknowns, including the composition variables, would exceed by even more the number of equations. If we are to find a dynamical recipe for unique fluid evolutions, we must find new kinds of equations, not new kinds of fluid mixtures. Going to mixtures is great for applications, but does not help us in our present search for the simplest fluid dynamical hints of determinism.

But before we give up on equations of change, let us look further. For example, let us consider the mechanical energy of our fluid piece as a property to be followed in time. Here the property of interest is the kinetic energy associated with the motion of our fluid piece as a whole, not that thermal or internal energy considered above. The rate of change of the property in one dimension can be expressed as

$$dx \frac{d}{dt}\left(\frac{m}{2} v^2\right) = \left[\frac{\partial}{\partial t}\left(\frac{\rho v^2}{2}\right) + \frac{\partial}{\partial x}\left(\frac{\rho v^2}{2} v\right)\right] dx \qquad (8.36)$$

and we can use (8.20), (8.27) to find the rest of the equation. But this adds nothing new to our recipe for predicting fluid evolutions, so we can omit it from further consideration. Here we bombed out.

Maybe we should go back to the expressions given above for the laws of thermodynamics. We have not examined those and there may be possibilities for dynamics there that we have overlooked.

To get at "internal energy" for the simplest kinds of particles, which only move, we compare their velocities with the average velocity (fluid convection) and average up the relative speeds, which are the speeds of the "internal" or "thermal" kinetic energy. The density of that relative kinetic energy is what we had in mind for the density u of internal energy. But u could include, of course, the energies that are "internal" to molecules, such as the various bond vibrational energies. We do not want to get into that here, so we restrict ourselves to monatomic molecules (inert gases). In that case the internal energy is directly related to temperature T (absolute). In fact, we could replace u in the formulas above by (3/2) nkT, i.e., by T as the variable of interest. This would emphasize the fact that u has nothing to do with the absolute velocity of our gas but only with the "spread" in relative velocities. Temperature is well known as a measure of that spread. For example, the temperature of a beam of molecules moving a million miles per hour in the same direction with no spread, capable of knocking down any resisting walls, is absolute zero. Temperature and internal energy have no correlation with the "directed" energy of a gas, but only with the "internal" (relative to directed) energy.

If we combine these relative translational energies with whatever other "internal" energies the molecules have, we can sum that to the internal energy U of the system, and the first law of thermodynamics expresses the change of U in terms of an input from heat dq and an input from mechanical work dw:

$$dU = dq + dw \tag{8.37}$$

where dq is energy into the body due to wall-collisions from outside that excite internal molecules to higher energies and dw is the contribution due to changes in the positions of containing walls due to the momentum gathered from molecular collisions.

So we have the first law of thermodynamics as applied to a body such as our fluid element, and expressed in a rate form, as

$$\frac{dU}{dt} = \frac{dq}{dt} + \frac{dw}{dt} \tag{8.38}$$

and we merely have to convert the notation into the form appropriate for fluid dynamics. For the left side of (8.38) we merely treat dU as the internal energy of our fluid element, which is the density u multiplied by the volume dxdydz, and we have, as we have had many times before,

$$\frac{dU}{dt} = dx\,dy\,dz\left[\frac{\partial}{\partial t}\,u + \frac{\partial}{\partial \underline{r}} \cdot (u\underline{v})\right] \tag{8.39}$$

In order to formulate the rate of accumulation of energy in our element due to heat conduction into and out from the element we have to introduce the concept of the heat flux vector q. In one dimension we may say that the rate of accumulation of heat due to the flux in at x and the flux out at x + dx is

$$q(x,t) - q(x+dx,t) = -\,dx\,\frac{\partial}{\partial x}\,q \tag{8.40}$$

Here, again, we have used the calculus of (8.11). Needless to say, the result (8.40) could be positive or negative depending on the actual values of q_x at the varuous x values. And we have not really done anything in introducing the heat flux except to add to our list of unknowns to be predicted.

In three dimensions, we have (8.40) as

$$\underline{q}(\underline{r},t) - \underline{q}(\underline{r}+d\underline{r},t) = -\,d\underline{r}\left(\frac{\partial}{\partial \underline{r}} \cdot \underline{q}\right) \tag{8.41}$$

which is easy to verify.

To formulate the work term in (8.37) we start in one dimension where the work is

$$dw = -\,p\,dL \tag{8.42}$$

where p is the force in Eq.(8.26) and dL is the change d(dx) in the length of our fluid element. So we have

$$\frac{dw}{dt} = -\,p\,\frac{d}{dt}\,(dx) = -\,dx\;p\left(\frac{\partial}{\partial x}\,v\right) \tag{8.43}$$

It may have seemed more appropriate to use

$$\frac{dw}{dt} = -\frac{d}{dt}(pdx) = -\frac{dp}{dt}dx - dx\,p\left(\frac{\partial}{\partial x}v\right) \qquad (8.44)$$

and that adds a new term, the first term on the right side of (8.44). But here we have to notice the subtle difference between exact differentials of a "state function" dU and the inexact nature of work (and heat). Since dw is not the differential of a function but rather the small increment of something along a specific path, we have to recognize (8.43) rather than (8.44) as the proper formulation for the "rate of work," better expressed as the "rate at which work on the system raises its energy U."

The generalization of (8.43) to three dimensions is complicated. There are contributions like the one described by (8.43) for each dimension (x,y,z) and there are contributions from the "shearing" components of the pressure tensor. They too exert a net force for the translation of our fluid element unless the shearing forces happened to be balanced, in which case they would change the shape of the fluid element without changing its velocity. If we considered all of the effects of the pressure tensor in detail, we would find a replacement for the one-dimensional (8.43) that looks like

$$\frac{dw}{dt} = -\,dxdydz\left[\underline{\underline{p}} : \frac{\partial}{\partial \underline{r}}\underline{v}_o\right] \qquad (8.45)$$

where, in our notation,

$$\underline{\underline{p}} : \frac{\partial}{\partial \underline{r}}\underline{v}_o = \sum_{ij} p_{ij}\frac{\partial}{\partial x_j}v_{oi} \qquad (8.46)$$

This includes not only the work of expansion (or compression) but also the work of shearing distortions due to viscous resistance.

Putting together (8.38), (8.39), (8.41), and (8.45) we have both the one-dimensional equation (8.4) and a three-dimensional equation for the conservation of energy:

$$\frac{\partial}{\partial t}u + \frac{\partial}{\partial \underline{r}}\cdot u\underline{v} = -\frac{\partial}{\partial \underline{r}}\cdot \underline{q} - \underline{\underline{p}} : \frac{\partial}{\partial \underline{r}}\underline{v}_o \qquad (8.47)$$

So now we have three general and always acceptable equations of change, which were listed (for one dimension) in (8.2) - (8.4). The problem is that they are three equations for five unknowns, and the situation is worse in three dimensions (five equations for 16 unknowns). We have to look further for more equations.

There is one and only one other place to look for a very general equation of change. That is the second law of thermodynamics. We are hesitant because of the common feeling that we might be leaving Newtonian dynamics, and hence all possibility of "understanding," maybe even waving in the air the kinds of statistical, quantumly mystical, cosmologically overwhelming arguments that have already done so much to obscure our objective. But there is a way to get into the second law without raising any flag. The fact is that there are two parts to the second law, and one of them contains no hint of irreversibility. It is that part, with no irreversibility, which we now consider, and no one will doubt our consistency with Newtonian molecular dynamics.

According to the second law, as we consider it here, there is a property to be called entropy and denoted by S such that dS for any reversible process (process at equilibrium) is dq/T. In other words, for a reversible process we have $dq = TdS$. We also have $dw = -p_e dV$ (in three dimensions), where p_e is the equilibrium (reversible) pressure. According to the first law, the same ultimate change, whether reversible or not, in U is the same

$$dq_{rev} + dw_{rev} = dU = dq + dw \tag{8.48}$$

And so we have

$$dq_{rev} = TdS = dq + (dw - dw_{rev}) \tag{8.49}$$

This equation does not say that dS must be positive for an isolated system or anything else that can be construed as an expression of any kind of irreversibility. Those familiar expressions of irreversibility have to be added, as if coming from another law. Equation (8.49) tells us how to calculate the entropy change for any process, including those with entropy changing the "wrong way," those which are actually excluded in the natural world.

And so we can accept (8.49) as the source of another very general equation of change which is not tied to irreversibility.

Dividing (8.49) by T, we have the rate of change of the left side in one dimension as seen by an observer moving with the fluid as

$$\frac{d}{dt} s = dx\left(\frac{\partial}{\partial t} s + \frac{\partial}{\partial x} sv\right) \tag{8.50}$$

where s is the density of entropy and sv is the convective flux. The first term on the right side of (8.49) offers no new problem, in view of our previous experience, and we have

$$\frac{1}{T}\frac{d}{dt} dq = - dx\left(\frac{1}{T}\frac{\partial}{\partial x} q\right) \tag{8.51}$$

Similarly, we have the rate of change of the second term on the right side of (8.49) as

$$\frac{1}{T}\frac{d}{dt} (dw - dw_{rev}) = - dx \frac{1}{T} (p - p_e) \frac{\partial}{\partial x} v \tag{8.52}$$

So we have the rate of (8.49) divided by T and dx as

$$\frac{\partial}{\partial t} s + \frac{\partial}{\partial x} sv = - \frac{1}{T}\frac{\partial}{\partial x} q - \frac{1}{T} (p - p_e) \frac{\partial}{\partial x} v \tag{8.53}$$

The difference between p and p_e is that p is the actual measurable pressure while p_e is the formula for equilibrium pressure in terms of other independent local variables (density and temperature). Here $\underline{q}$ is the heat flux.

If we derive the analog of (8.53) in three dimensions, we find

$$\frac{\partial}{\partial t} s + \underline{\nabla} \cdot s\underline{v} = - \frac{1}{T} \underline{\nabla} \cdot \underline{q} - \frac{1}{T} \underline{\underline{\tau}} : \frac{\partial}{\partial \underline{r}} \underline{v} \tag{8.54}$$

where the "stress" tensor $\underline{\underline{\tau}}$ is

$$\underline{\underline{\tau}} = \underline{\underline{p}} - p_e \underline{\underline{U}} \qquad \underline{\underline{U}} = \begin{pmatrix} 1 & 0 & 0 \\ 0 & 1 & 0 \\ 0 & 0 & 1 \end{pmatrix} \tag{8.55}$$

We notice that Eq.(8.53) added to the list of earlier equations of change in one dimension is an addition of one new equation that brings with it one new unknown, which does not bring us any closer to a dynamical scheme of equations. The same is true in three dimensions. No harm introduced by this entropy equation, but no gain either. We are still left with a set of equations of change which are obeyed by every Newtonian evolution of fluid behavior but selected by none. Fluids do not live by Newtonian equations of change alone.

In order to find a dynamical recipe for predicting fluid behavior we must now abandon our search for general equations that always hold and look for equations that hold under non-Newtonian restrictions. It is here that irreversibility has an opportunity to enter, if it must. We will certainly not accept any new restriction that contradicts or in any way goes against the Newtonian laws, for that would be tantamount to throwing in the whole towel. There is a modern tendency to throw in that towel. We object very strongly against that.

Now let us ask for the general features of the fundamental equations of change described above. One general feature is that they are not, by themselves, dynamical. They cannot offer a set of fluid equations that selects a particular fluid evolution based on initial conditions. They describe everything that can happen for a fluid composed of Newtonian molecules but they offer no clues as to why some things actually do happen and other things never happen. They seem to be necessarily incomplete, to ignore the demands of this world for irreversibility, to be as compatible with the processes that occur as with the reverses of those processes, which never occur. If we see them as determining they are an enigma. If we see them as simply incomplete, they are somewhat irrelevant.

Another feature of these fundamental equations of change, which simply adds to the emptiness described above, is that they are, by anyone's measure, invariant under "time reversal." They offer no hint of irreversibility in the sense of criterion 2. Look at each equation or its three-dimensional generalization, and replace t by -t, any motion variable like $\underline{v}$ or q by its negative, and you will recover the same equations. Those equations certainly cannot offer any selected direction of time. Reversing both direction and time leads to no change, which is to say that reversing direction <u>or</u> time leads to the same reversed sequence of events, which, incidentally, is familiar to us for having watched films run backwards.

8.5 Stochastic Equations of Change

If we identify equations of change as equations which contain time derivatives, as we have, then we have to notice that there are many common equations with time derivatives that come from purely probabilistic or intuitive reasoning. From our point of view the possibility of this kind of reasoning depends on an assumption of recent equilibrium, and the equations derived inherit irreversibility from that assumption. Let us examine the derivations of those equations.

First consider the case of the chemical reaction described by

$$2\ HI \rightarrow H_2 + I_2 \tag{8.56}$$

Visualize, if you wish, two molecules of HI colliding in such a way that the outcome can be two molecules H_2 and I_2. Not every such collision will produce those results, but there will be some collisions with enough energy and the right orientations to lead to those results. The other collisions produce nothing new, and we can forget them. So we may say that the rate of (8.56) is proportional to the collision frequency of HI molecules and to the probability of successful energies and orientations. This is all based on probabilistic reasoning, and it could be successful, unless, of course, some practical joker prepared a beam of HI molecules all headed in the same direction so that there would be no collisions. In other words there are conceivable Newtonian patterns of motion in which the probabilistic guesses would not apply. And yet, we assume, the guesses do apply in most situations, especially those situations in which the molecules have motions that are completely randomized, and the system has been isolated from jokers for some time.

We assume the motions of the molecules are randomized in the sense that the HI molecules are flying in all directions without any direction preference and have the equilibrium (most probable) distribution of speeds. And we assume that this state of affairs could continue indefinitely without the interference of unexpected "old" spontaneous withdrawals from randomness. This is the assumption of recent equilibrium. We cannot identify any other situation which could offer the disorder we need to proceed.

Under that assumption of disorder, we can make the following intuitive, probabilistic, stochastic assumption, which is widely known as the "mass action law." Under this law we may assume that the rate of the reaction (8.56) is proportional to the collision frequency which is proportional to the number density of HI molecules and again proportional to the density of collision targets, so it must be proportional to the square of the number density of HI:

$$\text{Rate} = -\frac{d}{dt}[\text{HI}] = k[\text{HI}]^2 \tag{8.57}$$

where [HI] is the concentration in moles per liter. Here we have used a traditional notation from the past, still current in chemical textbooks. If we use c = [HI], we have

$$\frac{dc}{dt} = -kc^2 \tag{8.58}$$

which can be integrated to find the unique and deterministic description

$$\frac{1}{c(t)} = \frac{1}{c(0)} + kt \tag{8.59}$$

If we rearrange the formula to read

$$c(t) = \frac{c(0)}{1 + c(0)kt}$$

we see explicitly the "fall off" of c in time leading eventually (asymptotically) to $c = 0$ after infinite time. Here we have a complete and dynamical account of the evolution of the system leading from its initial condition to the final condition.

To be more thorough within the framework of "chemical kinetics" we should include the reverse reaction

$$2\,\text{HI} \underset{k_2}{\overset{k_1}{\rightleftarrows}} H_2 + I_2 \tag{8.60}$$

and set up the problem of following the evolution as follows. Given the initial concentration $c = c(0)$ of HI, and assuming no H_2 or I_2 present initially, then we allow the reaction to proceed to the right until we have formed $x = x(t)$ moles per liter of H_2. The moles per liter formed of I_2 is necessarily equal to x, and the moles per liter of HI lost is necessarily 2x, from the stoichiometry of (8.60). Thus we have the rate equation as

$$\frac{d}{dt}[H_2] = k_1[HI]^2 - k_2[H_2][I_2] \tag{8.61}$$

which can be expressed as

$$\frac{dx}{dt} = k_1(c - 2x)^2 - k_2x^2 \tag{8.62}$$

The solution x(t) can be expressed implicitly by

$$at = \ell n\left[\left(\frac{bx - d}{bx - e}\right)\left(\frac{e}{d}\right)\right] \tag{8.63}$$

where

$$\begin{aligned} a &= 2\sqrt{k_1k_2}\,c \\ b &= 4k_1 - k_2 \\ d &= 4ck_1 + a \\ e &= 4ck_1 - a \end{aligned} \tag{8.64}$$

A graph of x vs t would show x(t) rising or falling monotonically and asymptotically to a level held indefinitely at equilibrium. The common idea of an irreversible process as an uninterrupted approach to an eternal equilibrium is very well illustrated by this behavior, as it was by (8.59) where the reverse reaction was missing. Including chemically reversed reactions or not, we find irreversible behavior which is dynamical and well described to the end.

There is a multitude of equations of change (rate equations) like (8.58), (8.62) governing evolutions from initial conditions based on intuitive, probabilistic ideas, including all theories about reaction rates, rates of nucleation, various relaxation phenomena, etc. And they are all dynamical and irreversible. They represent a modern approach to explanations of experimental results in time-dependent phenomena. They are the reason for the popularity of courses in "stochastic processes." And all for very good reasons. But in the end, we have to admit that all of those stochastic theories are possible only in a universe where recent equilibrium is the dominant predicament.

In the discussion above we have developed a sharp division between two different kinds of equations of change. The fundamental equations derived from body or fluid ideas, which are supported below by derivations from molecular motions, are completely reversible and general, covering all possible situations, but they are not dynamical, not complete in the sense of narrowing the fluid evolution to a particular path. Other laws must be added to provide a dynamical scheme, and those laws bring in the irreversibility associated with fluid behavior as observed. Those laws are merely stochastic, an insulting label, and yet they seem to be necessary as recipes or supplements to better recipes in order to provide dynamical schemes. In the case of the chemical reaction rates, the stochastic equations of change are self-sufficient rate equations that come from the "mass-action" law, which, in turn, comes from the assumption of recent equilibrium. If we considered reactions in a larger context including flow and conductions, we would need the fundamental equations of change as supplements, and maybe even some other equations. But we would need, in any case, at least some stochastic equations.

Our dilemma then is thus. We know that fluid behavior is irreversible in some sense and we would like to see the origin of the irreversibility appearing explicitly in the fundamental equations governing fluid behavior. Evidently, the possibility of such explicit signs of irreversibility is remote except in cases that include stochastic equations of change. We would prefer to see the signs in the fundamental equations which always apply rather than wait for stochastic equations that apply only in highly restricted circumstances, but that is not to be. Irreversibility seems to exist only in highly restricted circumstances. Maybe they are the prevailing circumstances, and that would explain irreversibility, but what are those circumstances and how do they prevail? Before we try to answer that, let us push on.

8.6 Simple Equations of Flux

Equations of flux have been discovered and developed for the ideas of conductive fluxes of mass, energy, momentum, and electrical charge. Attempts to find conductive fluxes of angular momentum and other molecular properties have not yet reached fruition. They probably never will.

We have equations for the fluxes suggested above, which come from the deep past:

Fick's Law of Mass Conduction

$$\text{flux of mass} = J_m = -D \frac{\partial}{\partial x} \rho \tag{8.65}$$

comes from the intuition of Fick as a physiologist, not especially interested in transport phenomena, but forced to face that in 1855 on the basis of observations of biological systems.

Fourier's Law governing heat conduction

$$J_{heat} = -k \frac{\partial}{\partial x} T \tag{8.66}$$

comes from the attempt by Fourier to find an example to illustrate his new technique in mathematics, now known as Fourier Analysis (1822). His intuition in selecting a model was, we would say, in retrospect, very good.

The law for charge conduction known as Ohm's law is commonly presented as

$$E = IR \tag{8.67}$$

where the flux I of current is the total flux through an area A, and we would describe it by a flux density J

$$J = I/A \tag{8.68}$$

Actually the force "field" E is the gradient of a potential (potential energy per unit charge) φ:

$$\underline{E} = -\underline{\nabla}\varphi \tag{8.69}$$

and the resistance R is the reciprocal of a "conductance" σA, so we have Ohm's law in the same pattern as the other transport laws,

$$J_{charge} = - \sigma \frac{\partial}{\partial x} \varphi \tag{8.70}$$

Ohm formulated this law in 1826 as an analog to the Fourier law for heat conduction. Here, at last, was a founding father especially interested in the phenomenon that bears his name. He also discovered "electromotive force" as the analog he searched for, analogous to Fourier's temperature difference (driving force).

In addition to these three flux laws, which work very well in many predicaments, there is a fourth flux law, which works well, except for some polymeric materials, known as Newton's law for viscous flow. Newton himself could not have cared more about viscous flow in real fluids than Fick could have cared about diffusion for its own sake or Fourier could have cared about thermal conduction for its own sake, but he did offer the idea which leads to a flux law for momentum conduction.

Consider a fluid moving in the y-direction with a fixed boundary at $z = 0$ so the velocity $v = v_y$ is 0 at $z = 0$ and increases with increasing distance z from the boundary, as shown in Fig.8.6. If the friction between fluid layers is

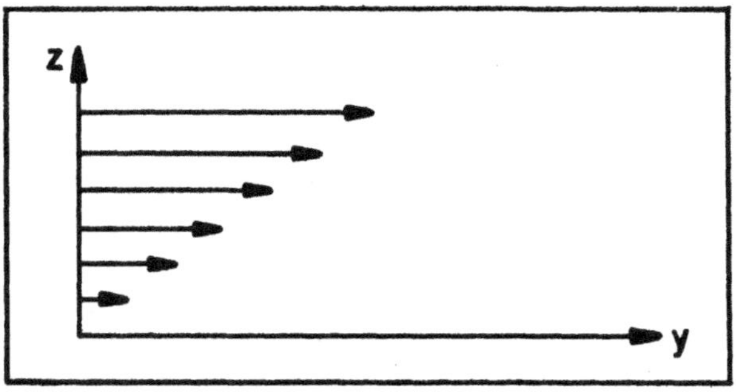

Fig.8.6 Velocity Profile in Viscous Fluid

large then the variation of the fluid velocity must be small, and if the friction is small the velocity variation can be large, as shown in Fig.8.7. Newton's idea was to take a given viscosity and say that the velocity variation would be large if there is a large shearing force (pressure) causing the variation. In fact, he proposed that the shearing force and the resulting velocity gradient are proportional

$$\tau = -\eta \frac{\partial}{\partial z} v_y \qquad (8.71)$$

The "stress" τ here is the force per unit area exerted from above on a z-plane in the y-direction, so it is what we have

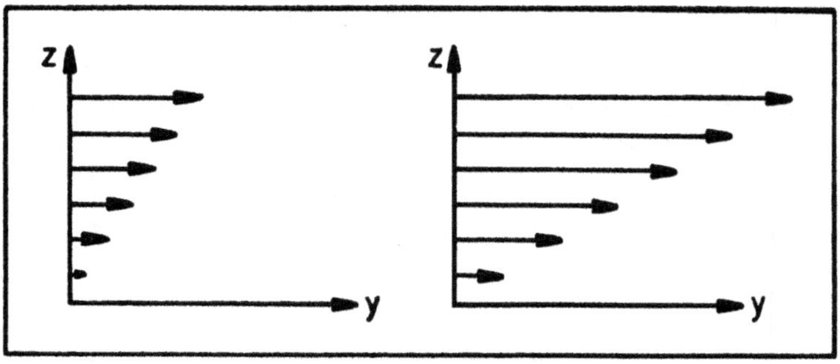

Fig.8.7 High Viscosity and Low Viscosity

previously referred to as the component p_{yz} or p_{zy} (the same),

$$p_{zy} = -\eta \frac{\partial}{\partial z} v_y \qquad (8.72)$$

This gives a formula for the off-diagonal elements of the pressure tensor and replaces those unknowns by formulas involving coefficients (constitutive parameters to be specified for every fluid) and velocity gradients. The number of unknowns can be reduced to the point where we have a dynamical system of equations. The system identified here with (8.72) is known as the Navier-Stokes system, which is widely used descriptions of viscous flow.

One familiar way to go from here is to assume that we have a steady state, and hence the time derivative is zero, and then to derive the Poiseuille law for the variation of v_y with the distance z from the axis of a cylinder through which the fluid is trying to flow. Other derivations are possible, of course. But the main ingredient in all such derivations, the ingredient which introduces irreversibility, is the flux law (8.71) or (8.72). That law is not compelled to hold but rather is assumed and is contingent upon the assumption of conditions where the shearing force can be proportional to the velocity gradient. Such a proportionality is not dictated by the molecular dynamics or by any other origin than the intuitive one. And it is observed that the law does not always hold for a variety of materials. So we may take Newton's law for viscosity the same way we take Fick's law for diffusion. They have the power to render dynamics when coupled with equations of change, and they render irreversible dynamics, but they also have the defect of being highly contingent and not generally reliable.

We now have a list of the simple equations of flux: (8.65), (8.66), (8.70), (8.72), and they all have the same form

$$J = L X \tag{8.73}$$

where J is the flux (conductive), X is a "force" (negative gradient of something), and L is the "transport coefficient" (diffusion coefficient or matter conductivity, thermal conductivity, electrical conductivity, viscosity or momentum conductivity). If there are other simple equations of flux, they have not yet been discovered. There was an attempt in recent times to find an equation of flux for angular momentum, but the macroscopic phenomenon seems to be missing.

Given these equations of flux we now face the questions: Are they dynamical? Are they reversible? How do they relate to the fundamental equations of change? First, on the question of whether the equations of flux are dynamical by themselves, the answer is clearly negative. Every one of these equations involves at least two unknowns. For example, even if we restrict ourselves to one dimension, looking at the diffusion equation [Eq.(8.65)], and even if we formulate the flux of mass as the mass density multiplied by the fluid velocity, we have (8.65) as

$$\rho v = - D \frac{\partial}{\partial x} \rho \tag{8.74}$$

which involves two unknown functions ρ and v. If we add to that equation the equation of change

$$\frac{\partial}{\partial t} \rho + \frac{\partial}{\partial x} \rho v = 0 \tag{8.20}$$

then we can eliminate v to find a dynamical equation

$$\frac{\partial}{\partial t} \rho = D \frac{\partial^2}{\partial x^2} \rho \tag{8.75}$$

which is known as Fick's "Second Law of Diffusion." Like any dynamical equation, this one is celebrated. The variety of evolutions it describes is well known.

Looking at the equation for heat conduction (8.66), where J in that equation is q in the one-dimensional form of (8.47), we have, in one dimension, and in the absence of fluid flow or gradients of any property other than temperature,

$$\frac{\partial u}{\partial t} = - \frac{\partial}{\partial x} q = k \frac{\partial^2}{\partial x^2} T \tag{8.76}$$

If we recognize the internal energy density u as the density of heat capacity c multiplied by the temperature T, we have (8.76) as a dynamical equation

$$\frac{\partial T}{\partial t} = \frac{k}{c} \frac{\partial^2}{\partial x^2} T \tag{8.77}$$

In fact, this dynamical equation has the same form as the diffusion equation (8.75) and its applications are as widespread.

We could find other dynamical equations by combining fundamental equations of change, which always hold, with stochastic equations of flux, which hold only under the most stringent conditions where intuition can be the guide. Evidently the flux equations supplement the fundamental equations and bring in determinism, i.e., dynamics. We could say the fundamental equations supplement the flux equations, but no one wants to refer to equations that always hold as "supplementary." So we will refer to the flux equations as supplementary.

Those supplementary flux equations bring in not only determinism and dynamics but also irreversibility. How they do that will be a question of interest below. At this point we may ponder the question of why the entry of irreversibility is accompanied by the entry of determinism. Are those two attributes of behavior somehow coupled, and if so, is that coupling an accident of nature or an ontological necessity of some kind?

The equations of flux were derived by applying probabilistic intuition similar to that behind the "mass-action law" of chemical kinetics needed for Section 8.5 above. That kind of reasoning has many labels which make it suspect: "intuitive, probabilistic, stochastic, heuristic," not to mention "mass-action-law-like" and "common sense." Why are those labels suspect? What is their common denominator?

What we have said before about that common denominator, and will say again here, and then say again in more detail below, is that the denominator points to a recent past equilibrium where the randomness justifies the suspect reasoning. The reasoning is logically justifiable at equilibrium, as we will see in the next chapter, and the only challenge can be about how that equilibrium gets into the considerations of post-equilibrium change.

Our conclusion at this point can be described as follows. We have the fundamental equations of change which are reversible in every sense, based on laws that have no foot in irreversibility and are known to be perfectly reliable as far as they go, and applicable to all fluid phenomena, and yet not deterministic (dynamical). We also have the equations of flux and they too, are not, by themselves dynamical. They are based on stochastic reasoning which was applied originally without any appeal to any kinds of fundamental laws, and they reek with the irreversibility we want to explore. In fact they reek so much that they led generations of students to look for irreversibility as a fundamental thing connected to some mysterious property of time itself, or to the stochastic nature of the description of phenomena as a reflection of God playing with dice.

We can, as pointed out above, combine fundamental equations of change with stochastic equations of flux to obtain successful, deterministic prescriptions of change which possess exclusion irreversibility. The origin of that irreversibility in the prescription is clearly buried somewhere in the equations of flux, not the equations of change. The origin of the success of those systems of equations is harder to assess. The only easy way to go here is to assume that the equations of change apply to a wide variety of phenomena, most of which never happen, and to see the equations of flux as a restriction that selects out a subset of phenomena that do occur, and occur with exclusion irreversibility. If we can identify that subset we may find the origin of exclusion irreversibility and the origin of determinism in fluid dynamics.

The search for those origins has been aided significantly by the method of statistical mechanics, which is described in the next chapter. Unfortunately, that search was accompanied by a continuing effort to find the solutions of origin problems within statistical mechanics itself, as if it were more than a tool. When we go to that method for help we have to be careful to avoid getting too much help.

8.7 Complex Equations of Flux

The equations of flux described above are "simple" in the sense that each kind of flux is caused by, and is proportional to, its own special and appropriate force (gradient of the appropriate "potential"). Thus a heat flux is due to, and proportional to, a temperature gradient. An electrical current (charge flux) is due to an electrical potential gradient (electrical force), etc. Who would have thought that a heat flux could be caused by an electrical force, or an electrical current could be caused by a temperature gradient? Intuition could never apply in such weird cases, if they occur?

And yet such weird cases do occur. It has been known since 1822 (Seebeck) that a temperature gradient can cause an electric current, and, in fact, the modern thermocouple for measuring temperature is an acknowledgement of that phenomenon.

And it has been known since 1834 (Peltier) that an electrical force (potential gradient) can cause heat flow. It is also possible for a temperature gradient to cause not only the expected heat flow in a gas, but also a diffusion flow (Soret effect), which is expected to be caused by a concentration gradient, and for diffusion to be caused by a temperature gradient (Dufour effect). Evidently there is a "coupling" between different kinds of transport processes. In fact, it has become necessary to generalize the simple flux laws of the previous section,

$$J_i = L_i X_i \tag{8.78}$$

to a linear combination

$$J_i = \sum_j L_{ij} X_j \tag{8.79}$$

These are the "complex" equations of flux.

The fact that these complex equations are linear is not, by itself, an alien to intuition. Once we know that there is a contribution J_1 to the heat flux due to the force X_2 (gradient of concentration) that is expected to cause only diffusion (not heat flow), then we can easily believe that J_1 is proportional to X_2,

$$J_1 = L_{12}X_2 \tag{8.80}$$

According to intuition, "effects" are always proportional, or nearly so, to "causes."

What is not intuitive about such a coupling between the cause for diffusion and the effect in heat flow is that there should be any such coupling at all. So this coupling offers a new clue about the origins of the irreversible flux equations. We noticed above that the proportionality in simple flux equations (8.78) pointed to recent equilibrium as its origin, but here we have a different kind of clue, and it remains to be seen whether it points to recent equilibrium or some other source.

Our earlier suggestion above about the simple flux equations selecting out a subset of Newtonian phenomena that actually occur in fluid behavior is now complicated by the question of whether that same selection could happen to be the one that allows for couplings of transport processes with the reciprocal relations (4.3). We could be putting too much pressure on the idea of recent equilibrium as the origin of fluid behavior.

On the other hand, if it turns out that recent equilibrium is the determining factor in the predicament of observed fluids, then we have severe irony in the terminology surrounding complex equations of flux. The Onsager explanation (1931) of his reciprocal relations has been interpreted as an explanation of how <u>macroscopic irreversibility</u> can arise from <u>microscopic reversibility</u>, leading to an adulation of <u>microscopic reversibility</u> (Note 27). But it may be that <u>macroscopic reversibility</u> is the origin of the reciprocal relations, and the problem is to explain how that comes out of <u>microscopic irreversibility</u>. Of course we are mixing here various meanings of "reversibility," but not in any way that is wrong. References to reversibility and irreversibility in discussions about the reciprocal relations are references to temporal reversibility (criterion 2) and they are wrong because there is no such thing as temporal irreversibility. Our references to reversibility were references to inertial reversibility, where we really do have reversibility at the macroscopic level emerging naturally from irreversibility at the microscopic level (fluid equilibrium from molecular motion). The problem here is not merely one of semantics but one that is deeply rooted in the "problem of irreversibility."

We defer to the next chapter, which is on statistical mechanics, any further consideration of the complex flux laws. There are many interesting applications which could be considered, but they shed no light on the question of interest here. The question of interest there will be this. Given that statistical mechanics can lengthen our intuition while not going against mechanics, can it lengthen our intuition enough to understand the occurrence of coupled transport processes and explain the reciprocal relations?

8.8 Equations of Equilibrium

These are the majority of equations that fill textbooks on science and engineering at all levels, starting in the ninth grade. If for no other reason than that, we should wonder about the primacy of equilibrium predicaments in our considerations of fluid behavior. From the point of view of fluid dynamics and pedagogy, these equations are the simplest equations and they apply to very special cases. From our point of view, the equations of equilibrium may be simple in form and they may be specializations of other equations in some cases, but they describe the most sophisticated models and the most common phenomena. And they are so full of irreversibility that they are "saturated" and "neutral," offering no explicit hints about the nature of irreversibility.

After that somewhat cryptic introduction, let us try to illustrate the equations and organize an overview. Then we will ask for examples to show how the equations are needed to complete dynamical schemes and ask for the implications about the origin of irreversibility in fluid dynamics.

The equations of equilibrium can be grouped into various categories that reflect common denominators, e.g., common concepts, common devices for measurement, and common equations of equilibrium. For example, consider the "thermal" properties of matter. These are related to "heat" and they include properties that can be measured with a thermometer and a calorimeter. The thermal properties of a material include special temperatures, such as melting and boiling points, heats of melting and boiling, and heat capacities. If we put a heat q into a material and it does not melt or boil but rather its temperature rises by ΔT, then we may guess that the heat is proportional to the temperature rise

$$q = C\Delta T \tag{8.81}$$

where the proportionality coefficient C is the "heat capacity." If C varies significantly with T then we have to express (8.81) as

$$q = \int_{T_1}^{T_2} C(T)\, dT \tag{8.82}$$

Experimental formulas for C(T) are sometimes called "thermal equations of state." (In the various theories about equilibria, equations are often referred to as "equations of state.") In the case of (8.82) the equation is regarded as an equilibrium equation because it is assumed that the heat is passed inwards slowly and "reversibly" allowing the system to adapt instantaneously and always be at equilibrium, or at least, "local equilibrium."

Another category of properties gathers together the "mechanical" properties of a material. For an ideal gas the mechanical properties are all contained in a simple, familiar, mechanical equation of state:

$$pV = nRT \tag{8.83}$$

More complicated equations are available for other models. Generally, for liquids and solids, the equation is not available and we have tables of values for "compressibilities" or "expansivities" (at fixed pressure), etc. All of these properties are equilibrium properties in the same sense that (8.83) is an equilibrium equation. It is assumed that there is no temporal change, no spatial variation of p, n, and T; the gas has settled down to a uniform and constant future. We may use (8.83) or any other such equation of state away from equilibrium, but always with some uncertainty depending on how far things have changed.

There are many other categories of equilibrium properties and equations of equilibrium. Solution properties include solubilities, vapor pressures, ... and how they vary with temperature, maybe even pressure. Chemical properties include equilibrium constants and how they vary. Electrochemical properties include cell voltages and how they vary ... . We could go on somewhat indefinitely, and even include molecular properties as revealed by spectra taken on materials at equilibrium disturbed by the incident radiation. Each category has its own data and "equations of state," i.e., equations of equilibrium.

And we need not end the list of categories without considering phenomena controlled by external forces or lower dimensional phenomena. The effect of gravity causing the atmosphere to be denser nearer the earth, the effect of a centrifuge causing heavier particles to be further out in the tube at equilibrium, the effect of the surface tension in a bubble or a balloon causing the inside pressure to be greater by just the right amount than the external pressure. The equilibrium involved in the stretch of a rubber band, the equilibria involved in geological models involving stresses and strains and diffusion slow enough to be regarded as reversible on our time scale. There is almost no end to the list of categories of equations of equilibrium. The world is surely dominated by equilibria.

The most remarkable source of equations of equilibrium is Thermodynamics. This chapter of the science of equilibrium properties of matter is unique in many ways. First of all, it has no properties of its own. It is not simply another one in the list above. But rather thermodynamics is the central "phone company" that connects all the other chapters to each other, making it possible to derive data and equations in one category from data and equations in another. This "cross-linking" which is offered by thermodynamics provides the most spectacular equations of equilibrium. The role of thermodynamics as the super source of equations of equilibrium is illustrated in Fig.8.8.

Each of the areas connected to thermodynamics in Fig.8.8 has its own data and equations of equilibrium (equations of state) but depends on the mathematical recipes of thermodynamics for connections with other areas, leading to an almost unlimited supply of equilibrium equations. We can always connect a new area of data and equations into the the thermodynamic "tunnel" and reap harvest upon harvest for our effort.

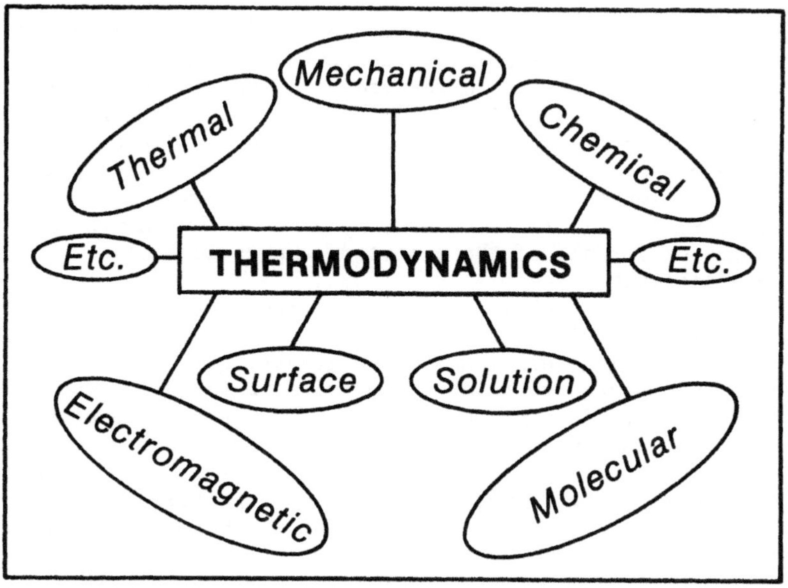

Fig.8.8 The Role of Thermodynamics

Incidentally, the linkage between molecular properties and the thermodynamic tunnel is the method known as "statistical mechanics." That method can be generalized to nonequilibrium situations where the linkage is to the "thermodynamics of irreversible processes," but the generalization has to be carried out very carefully. Strictly speaking, statistical mechanics is a method for describing equilibrium situations, and the extension must be guided by the very things we are looking for here, origins of irreversibility.

But getting back to equilibrium, and looking again at Fig.8.8, we notice first of all that thermodynamics plays a central and synthesizing role in the theory of fluid equilibria. That implies that we have to pay good attention to thermodynamics if we want to have an overview of equations of equilibrium. But that does not necessarily mean that we have to look for irreversibility in thermodynamics. The irreversibility could be located separately in each of the areas connected to thermodynamics, and it need never appear within the thermodynamic tunnel.

What actually happens is that irreversibility does appear in each of the areas connected to thermodynamics and does have its general reflection within the tunnel, and that can be bounced back out to the areas to add new information coming from irreversibility. In fact, the famous version of the "second law" of thermodynamics which says that the entropy of the world increases is a summary of what happens within the tunnel when we are considering irreversible processes. The increase of entropy becomes the common denominator for all of those outside expressions of irreversibility.

The statement above that equilibrium thermodynamics has no properties of its own should be reconsidered. It does have its "helping functions" which greatly simplify the derivations needed to relate properties in the different outside areas. It has, for example, the "internal energy U, enthalpy H, entropy S, free energies A and G," not to mention the "chemical potential" μ and the "activities" and "activity coefficients," etc. And some of these properties can be interpreted in terms of concepts prior to thermodynamics. The internal energy U, which is mysterious at the continuum level, is actually the Newtonian kinetic and potential energies of the molecules that make up the fluid. And the entropy is known to be related by $S = k \ln W$ to the number of ways W the molecules can make up the fluid. The other "thermodynamic properties" have no such secure interpretations. But whether or not these properties are interpretable, they are not the same kinds of properties as the properties in the outside areas mentioned above. They are not wanted for themselves and they are not directly measurable or controllable. They are "intermediates" in derivations of equations of equilibrium which finally involve only the properties of the "outside" categories. One could argue, therefore, that the "machinations" of thermodynamics, leading to titles of "machinery" for chapters in recent textbooks (Atkins, 1980), are not so much equations of equilibrium as they are simply manipulations of "helping functions," i.e., pure mathematics. But those machinations do often produce equations of equilibrium like

$$\frac{d \ln x}{dT} = - \frac{q}{RT^2} \tag{8.84}$$

where x is an equilibrium property of interest such as vapor pressure, solubility, equilibrium constant, ... and q is a corresponding heat. These are the most sophisticated of our equations of equilibrium and they connect properties of different categories explicitly revealing the role of thermodynamics. The point to be made here, if not clear, is stated much better on the first page of the remarkable book by Denbigh (1971).

The remarkable feature of equations like (8.84) is they relate macroscopic properties from various categories outside of thermodynamics in Fig.8.8 without any necessary reference to the thermodynamic intermediates used to derive the equations. Those intermediates have done their job and have been retired when we arrive at (8.84). The "intermediate," "helping" equations in the middle of the thermodynamic "tunnel" are equations of equilibrium, and there is a million of them, so we may say that equations of equilibrium are much more numerous than any other kinds of equations. This may not be a fair way to compare the numbers of equations since the thermodynamic equations are merely "helping" equations for derivations. Equations of flux have their helping equations but they still do not have a fair chance in this competition because the Irreversible Thermodynamics is a very small theory with a very small task compared to the Equilibrium Thermodynamics. And the voice of thermodynamics is shrunk to zero in the theories about equations of change, except insofar as those theories can borrow from thermodynamics with hopes for guidance and approximations. They rely on the helping functions in other chapters of physics, such as the potential energy of mechanics and the electrical potentials, but those nonthermodynamic helping functions are available for applications to fluid equations only in their own area of application. They are not thermodynamic.

In a way it is foolish to count up the numbers of equations of the various types. But there are clues in that record that tell us something about irreversibility. As we move from fundamental equations of change, with no irreversibility, to stochastic equations of change and flux, which are irreversible, to equations of equilibrium, which are saturated with irreversibility to the level where nothing can change, we find an increasing dependence on thermodynamics. And thermodynamics itself describes systems which are in their most probable conditions as determined by the states of the underlying molecules and the macroscopic constraints on the choices of states. There must be a connection between irreversibility and probability even though it cannot, of course, be a causal connection.

Why are there so many equations of equilibrium? And why is thermodynamics so important, so widely applicable? Its scope is even broader than indicated above. Many of the equations of mechanics for fluids can be derived from thermodynamics. Biology and Geology are undergoing revolutions due partly to the intrusion of thermodynamics. Ideas from thermodynamics are beginning to be applied in Economics and Sociology, and in applications of Statistics that have no direct connection to fluid dynamics (Information Theory).

For fluids the answer may be that equilibrium is the predominant condition encountered in nature. These fluids are either at equilibrium or else in a condition of having been recently at equilibrium and forced to withdraw at some recent known time in the past. Their equilibrium history still dominates their behavior. If fluxes have developed they are still predictable from the past equilibrium. If "relaxations" have developed, they are still predictable from the past equilibrium. And we always have the fundamental equations of change irrespective of the history if they are needed. The fundamental equations by themselves do not form dynamical systems, so we always have to reach for irreversible equations to obtain a dynamical prediction. By "irreversible" equations we mean equations with the right recent history which excludes the reverses of observed processes. The question of how that exclusion follows from the history of recent equilibrium was considered earlier and we will return to it later. The one clear conclusion we can draw at this point is that observed fluids have a great tenacity for equilibrium.

The case of the living organism is different. They thrive on interruptions caused by outside changes (nourishments, distractions, breathings, etc.) so frequent that the system is never allowed to reach total equilibrium (death). And we can assure the needed frequency in additional cases by offering resuscitation. But eventually there is equilibrium, and no way to stave it off indefinitely. Equilibrium always prevails.

Let us now digress to notice how equations of equilibrium can be combined with other equations to form dynamical schemes. The forming of dynamical schemes has been the theme of this chapter, and we want to bring that theme to completion. We have encountered the more important theme concerning the growing importance of irreversibility as we move from equations of change to equations of equilibrium, but nevertheless we return to the original theme.

The example we choose involves an application of the first law of thermodynamics. According to the first law there is a function U (internal energy) such that for any process in a simple system, dU is the sum of the heat into the system and the work on the system,

$$dU = dq + dw \tag{8.85}$$

where the work is

$$dw = - p_{res} dV \tag{8.86}$$

For a reversible process (reversible in the sense of criterion 1) the resisting pressure is equal to the pressure of the system p, and for an ideal gas this is

$$p = (nRT)/V \tag{8.87}$$

We also have U as a function of any independent variables x and y for the system and if we choose x and y to be T and V we have the "total differential"

$$dU = \left(\frac{\partial U}{\partial T}\right)_V dT + \left(\frac{\partial U}{\partial V}\right)_T dV \tag{8.88}$$

If $dV = 0$, then dU is the first term on the right of (8.88) and at the same time it is the heat (no work at constant volume) dq_v, which is experimentally given as $nC_v dT$, where n is the number of moles and C_v is the heat capacity at constant volume. Thus we have (8.88) as

$$dU = nC_v dT + \left(\frac{\partial U}{\partial V}\right)_T dV \tag{8.89}$$

Let us now restrict ourselves to ideal gases in adiabatic processes. This is the problem of "refrigeration" in which gases (nearly ideal) expand in an insulated environment where heat cannot come in or go out. It is also the problem of sound-wave propagation or the propagation of any kinds of waves through a gas (nearly ideal at ordinary pressures) if the waves are not so "big" as to drive the system too far from equilibrium. The wave propagation involves a sequence of rarefactions and compressions that occur so fast that there is no time for heat flow. Hence the process turns out to be adiabatic with dq = 0. Then we have (8.85) - (8.87) as

$$dU = dw = -\,pdV = -\,(nRT/V)\,dV \tag{8.90}$$

At the same time we have (8.89) for any ideal gas process in the form

$$dU = nC_v dT \tag{8.91}$$

The absence of the second term on the right side of (8.89) in (8.91) has nothing to do with the process being adiabatic, but is due solely to the fact that the gas is ideal. In an ideal gas there are no interactions between molecules (no potential energy, just kinetic energy) and hence the energy U cannot depend on how far apart the molecules are. It cannot depend on V. Combining (8.90) and (8.91), which hold separately for completely different reasons, and eliminating dU (retiring U as a good and faithful servant), we have the remarkable equation

$$nC_v dT = -\,nRT\,\frac{dV}{V} \tag{8.92}$$

We can generally assume that C_v is constant. Thus dividing (8.92) by T yields an equation which can be integrated

$$C_v\,\frac{dT}{T} = -\,R\,\frac{dV}{V} \tag{8.93}$$

to become

$$C_v\,\ell n\,\frac{T_2}{T_1} = -\,R\,\ell n\,\frac{V_2}{V_1} \tag{8.94}$$

The use of thermodynamics has been completed and we have (8.94) standing there waiting for application. One application is to ask what happens if we expand a gas adiabatically from an initial volume and temperature V_1 and T_1 to a larger volume V_2. How far will the temperature fall? What is T_2? We can solve (8.94) to find T_2.

The application to sound waves involves using pV = nRT to change (8.94) into the more familiar form

$$p_1V_1^\gamma = p_2V_2^\gamma \qquad \gamma = \frac{C_v + R}{C_v} \tag{8.95}$$

We could just as well say that pV^γ is constant. To convert (8.95) to its fluid form, we notice that for a particular sample of the fluid, the volume is inversely proportional to the density ρ. And so we have

$$p\rho^{-\gamma} = C \tag{8.96}$$

where C is a constant that can be evaluated as the left side of (8.96) where p is the average p and ρ is the average ρ.

Now let us take stock of what we have for fluid dynamics in the case of the adiabatic propagation of sound waves in an ideal gas. We could do this with equal ease for the one-dimensional gas and the three-dimensional gas, but we go for the one-dimensional gas to simplify notation. We have the equation of mass conservation (equation of "continuity") (8.2), which always holds but which is inadequate by itself because it is one equation for two unknowns ρ and v. We add to that the equation of motion (8.3) which always holds but brings in a new unknown p. (Ignore the term involving the external force.) So we have two equations for three unknowns ρ, v, p. We need a third equation which does not bring in any new unknowns. That equation is (8.96). Then we have a dynamical scheme and we can use that to describe sound propagation. The common result is a calculation of the speed of the propagation and how that depends on the average p and the value of γ. (See any text on fluid dynamics.)

This application of fluid equations involved two fundamental equations of change and one equation of equilibrium. We previously considered an application that involved one fundamental equation of change and one equation of flux leading to Fick's dynamical equation (8.75). Needless to say, there are many other applications of the fluid equations which involve mixes of equations of the various types brought together to form a dynamical scheme. In every case there is the contamination by the virus of irreversibility, the symptoms of equilibrium or its recent presence. If we stay with the "pure" and reversible, fundamental equations of change we never get a dynamical scheme, we never predict anything, we never reach the "real world." We must always include equations of flux, which point to a recent equilibrium, and/or equations of equilibrium, which emphasize even more that recent equilibrium.

Our conclusion for this chapter is that the variety of fluid equations needed to describe any actual fluid phenomenon may or may not include fundamental equations of change but must always include stochastic equations of flux and/or equilibrium. The irreversibility of the phenomenon itself, insofar as it can be described by equations, must come somehow from equilibrium.

CHAPTER 9

IRREVERSIBILITY IN STATISTICAL MECHANICS

The method of statistical mechanics has been referred to many times in earlier chapters without consideration for the details of how it works. Some of those details are helpful in understanding the clues about irreversibility that come from statistical mechanics. In this chapter we try to provide the needed details for the case where the underlying mechanical model is classical (strictly Newtonian).

In the first three sections we review the method of statistical mechanics as applied to equilibrium situations. We start with the application to an ideal gas at equilibrium. At first sight this seems to be a digression to a case that is trivial in a general discussion about fluid behavior. But the method of statistical mechanics is fully illustrated by this application. We will find that applications to systems with interactions and systems leaving equilibrium require the framework of the method as presented in the first section.

In the fourth section we depart from statistical mechanics and express Newtonian mechanics in the form needed for a systematic joining to statistical mechanics. We describe the Newtonian evolution of a probability fluid by assuming that each point in the fluid follows a Newtonian trajectory. Finding the probability of any point as an initial condition is a problem for statistical mechanics, and then following the evolution of that probability is a problem for mechanics. There is a clean separation of roles for statistics and mechanics here, and the joining of them involves no conflicts. We will introduce no stochastic obstacles (e.g., mass action law, molecular chaos). When we come to the final synthesis we will have a method which has no flaws except that it is not deterministic. To proceed further toward predictions of fluid behavior it will be necessary to introduce nonmechanical assumptions to bring determinism into the scheme. Those assumptions will also bring in irreversibility, and that is why we want to know what those assumptions are.

The introduction of an extraneous assumption to bring the scheme into a predictive mode will be described in the fifth section. Attention there is given to cases where we can ignore the interparticle interactions (rarefied gas) or else treat the interactions in an averaged way (mean-free-path theories for dilute-but-collision-dominated gases).

In Sections 9.7, 9.8 we describe the bold step toward gases with interactions taken by Boltzmann (1872), advanced by Enskog (1922), and revived in recent decades under the label of the BBGKY theory described in Section 4.4. The issue of main interest will be, again, the extraneous assumption that must, or seemingly must, be introduced to make the scheme predictive, and, inevitably, irreversible.

9.1 The Method of Statistical Mechanics

The mechanical states of one individual particle are identified by values of certain variables such as position x and velocity v for a particle in one dimension. All other properties of that particle can then be formulated in terms of those basic independent state variables. For example, kinetic and potential energies can be expressed in terms of position and velocity.

In three dimensions we would say that the state $\underline{s}$ is a collection of six property values:

$$\underline{s} = (\underline{r},\underline{v}) \qquad \underline{v} = (v_x,v_y,v_z)$$

$$\underline{r} = (x,y,z) \qquad \underline{s} = (x,y,z,v_x,v_y,v_z) \qquad (9.1)$$

For a particle in a quantum mechanical model, we would use the quantum numbers to provide the state address. In either case, if we know the state of the particle then we can find from available formulas the values of all properties of interest, including, of course, the original state variables themselves. For example, if we know $\underline{s}$ in (9.1), then we have

x-coordinate of position = x

x-component of velocity = v_x

y-component of momentum = mv_y

kinetic energy = $\frac{1}{2}m(v_x^2+v_y^2+v_z^2)$

z-component of angular momentum = $m(xv_y - yv_x)$

etc.

The essential feature of a state designation $\underline{s}$ is that all properties can be calculated by formulas from $\underline{s}$.

But suppose we do not want to get into the details about selections of state variables and formulas for properties. We want to defer that to the end of the line and concentrate here on the statistics. Then we could merely enumerate the states by a label $s = 1, 2, 3, 4, \ldots, k$, where k is the total number of states available for one particle (k can, of course, be infinite). This is a kind of "quantization for convenience" or a severe abbreviation, but it does help.

Next we consider a collection of n particles, each of which is in some one-particle state s. We may say that particle i is in state s_i, where s_i can be (using the label idea for s) $s_i = 1, 2, 3, \ldots$. In fact we can identify the super state of the whole system of n particles by

$$\underline{s} = (s_1, s_2, s_3, \ldots, s_n) \tag{9.2}$$

We could, for example, have nine particles distributed over four one-particle states, and one of the super states is

$$\underline{s} = (1, 4, 1, 1, 2, 1, 2, 3, 4) \tag{9.3}$$

Notice that in this super state we have four particles in state 1, two particles in state 2, one in state 3, and two in state 4. Thus we can describe the state in less detail (not saying what particle is in what state) by merely specifying the "distribution" $\underline{n} = (n_1, n_2, n_3, n_4)$ where n_s is the number of particles in state s. For the case of (9.3) we have

$$\underline{n} = (4, 2, 1, 2) \tag{9.4}$$

The early but crucial step in the method of statistical mechanics is to recognize the difference between the detailed designation of a super state, as in (9.3), and the less detailed designation of the distribution in (9.4). Macroscopic properties depend only on the distribution. They could not care less what particular particle is in what state, but they do depend on how many particles are in each state.

Consider, for example, a molecular property $\underline{a}$ which is additive in the sense that the macroscopic value A is a sum of contributions from individual molecules. This is the case for kinetic energy, momentum, angular momentum, and others. Summing over the molecules, we have

$$A = \sum_i a_{s_i} \tag{9.5}$$

where the sum is over molecules numbered by $i = 1, 2, 3, \ldots$. This formula is useless. We will never know the states of individual molecules, and if we did we could never perform the summation (9.5) over Avagadro's number of molecules (10^{23}). Using a superstate designation $\underline{s}$ to derive macroscopic property values from molecular property values is hopeless. And furthermore, it is not clear why we would ever want to do that, for we never know what superstate the system (fluid) is in. But if we move away from superstates to distributions, we find a different story.

First of all, having moved to distributions, we find the addition formula (9.5) simplified from a sum over states of individual particles to a sum over the populations n_s of the states s:

$$A = \sum_s n_s a_s \tag{9.6}$$

This is a simpler sum than the one in (9.5) if we have a way of knowing the populations n_s, i.e., the distribution $\underline{n}$,

$$\underline{n} = (n_1, n_2, n_3, \ldots, n_k) \tag{9.7}$$

In general we have no more way of knowing a distribution $\underline{n}$ than a superstate $\underline{s}$. But there is a case where we can know $\underline{n}$ better than we know $\underline{s}$, and know it well enough to calculate macroscopic properties A from molecular properties a_s in the manner prescribed by (9.6).

So the second reason for moving to distributions is that we can know more about the distribution for a macroscopic predicament than we could ever know about the state $\underline{s}$. This comes about from the fact that the probability of one distribution can be nearly infinitely greater than the probabilities of other distributions leading us to nearly certain knowledge of the distribution prevailing in an actual gas even though we may never know anything about the state $\underline{s}$ of the gas. At first sight, this may seem incredible, knowing so much about the distribution without having prior knowledge about the "full-fledged state" (superstate) s. But that is what happens, and the way it happens is the method of statistical mechanics, which we will review below.

Let us admit from the outset that this talk about probabilities is surprising in a discussion about dynamics. If we are after dynamics, which predicts exactly which way something goes, we do not want probabilities that the direction was this or that. That would not answer our question. But suppose you find all those probabilities for all of those directions and it turns out that the probability of one direction is infinitely greater than the probabilities for other directions. Then that would lead to the selection of that infinitely probable direction as the one and only direction, predicted with dynamical certainty. That is essentially what happens in predictions by the method of statistical mechanics. Our reference to "infinitely greater probability" is a bit overdone, but the prejudice is so great that you would have to measure properties to much greater precision (a dozen significant figures) than is possible or interesting to begin to see "fluctuations" (errors) in the predictions of statistical mechanics.

Now let us consider the method itself and look for the reason it is so successful. We will start with talk about probabilities but that will soon evolve into talk about certainties. To begin, we refer to the superstates s and ask for the probability of each s. The only answer that can be given to that question is that all s are equally likely. That is regarded as the fundamental assumption of statistical mechanics (Fowler and Guggenheim, 1939, Chapter 2). It is also the expression of the modern view (Poincaré) of probability theorists about the deepest interpretation of probability. If the assumption of equally likely points in the first sample space is wrong, then there is a "loading" of states that points to a more fundamental sample space. And the equality of probabilities for various fundamental states s agrees with intuition. Physics would be lost without that. On the other hand, we do have to qualify the assumption in the same manner. Two possible states s are equally likely if both are "allowed," i.e., consistent with the given data for the state of the system. In a way, it goes without saying that any state s which is impossible in view of the given data must be assigned a probability value of zero. For example, if the volume of the system described by $\underline{s}$ is bigger than the volume of our macroscopic sample, then the probability of $\underline{s}$ must be zero. So we should be referring to "accessible" (allowable, consistent with given data) states and only those when we say they are equally likely. This restriction, which may be obvious, is explained very thoroughly by Fowler and Guggenheim (1939).

If all the accessible states $\underline{s}$ are equally likely, then the probability of each is 1/N, where N is the total number of such states. So what is the probability of a particular accessible distribution $\underline{n}$. That is the crucial question. The answer is simple enough. The probability is 1/N multiplied by W, where W is the number of states $\underline{s}$ with the distribution $\underline{n}$.

For the reader who is not familiar with these arguments, let us pause here to consider an example. Suppose our system of n particles is three coins, $n = 3$, each with two states $s = 1$ (heads) and 2 (tails). The possible states are listed in the left-most column of Table 9.1. The next column lists their $\underline{s}$ designations. The probability of each state is 1/8. The third column lists distributions $\underline{n}$ and the arrows to that from the second column show the many-to-one correspondence between $\underline{s}$ and $\underline{n}$. The next column displays values of W, which specifies how many $\underline{s}$ correspond to each $\underline{n}$, and the last column shows the probabilities of the various distributions $\underline{n}$.

TABLE 9.1

ILLUSTRATING THE ROLE OF W FOR THREE COINS

States	$\underline{s}$	$\underline{n}$	$W(\underline{n})$	$p(\underline{n})$
HHH	(1,1,1) ←	(3,0)	1	1(1/8)
HHT	(1,1,2) ←			
HTH	(1,2,1) ←	(2,1)	3	3(1/8)
THH	(2,1,1) ←			
HTT	(1,2,2) ←			
THT	(2,1,2) ←	(1,2)	3	3(1/8)
TTH	(2,2,1) ←			
TTT	(2,2,2) ←	(0,3)	1	1(1/8)

The thing to notice is that the probabilities $p(\underline{n})$ of the distributions are proportional to W. If we do not care about precise (normalized) probabilities but only about the relative probability of one outcome relative to another, we could follow W and forget $p(\underline{n})$. That is our preference, incidently, right up to the point where the inscription on Boltzmann's tombstone is $S = k \ell n W$, expressing thermodynamic entropy in terms of W rather than $p(\underline{n})$. We never need the added detail of $p(\underline{n})$.

The point to be made is that there is a function of $\underline{n}$, namely W, which measures the relative probability of various distributions and can identify an unusually probable distribution by being very much larger in value than W evaluated for other distributions. In the example above (Table 9.1), W is larger for distributions (2,1) and (1,2) than for (3,0) and (0,3), and we would bet fifty cents that a flipping of three coins would give those more probable distributions. But the more probable distributions are not that much more probable as to warrant a bet of a week's pay. What we are doing here is applying statistics at its lowest level, and getting nothing much from that.

But if we go to Avagadro's number of coins (6.02×10^{23}), which is the typical number of molecules in an observable sample, then W becomes a function which is so strong for the most probable distribution (largest W) that we could bet a year's pay on that distribution and be assured of winning. By "that distribution" we mean something like between 49.99999% heads and 50.00001% heads. When the number n becomes large, the most probable distribution becomes more and more probable. That is a statistical thing that has nothing to do with the dynamics of coin flipping.

What is needed here to confirm the speculations above and to provide a method for calculations in other problems is a formula for W. The formula is well known and should be on the front cover of any text on statistical mechanics:

$$W = \frac{n!}{n_1!\, n_2!\, n_3! \ldots\ldots n_k!} \tag{9.8}$$

This formula gives the number of ways to distribute n things into k baskets and get n_1 in the first basket, n_2 in the second, etc. It is also the number of superstates $\underline{s}$ that have the distribution $\underline{n}$. Anyone who wants to thoroughly understand the method of statistical mechanics will have to develop some examples and become convinced that (9.8) is the right formula for W.

Now that we have the formula for $W(\underline{n})$ for the relative probability of the distributions we can find the most probable distribution by looking for the one that maximizes W. Our search can be treated as a problem in the calculus of variations. Let us forget, for the moment, that the n_s are integers and treat them as continuous variables. Then we can start with any distribution $\underline{n} = (n_1, n_2, \ldots)$ and acknowledge its value of W and then let it vary a little bit where each n_s moves to $n_s + dn_s$ and then determine whether those variations cause an increase or decrease in W. Whatever the change dW in W we can find it from the changes dn_s by using the formula

$$dW = \sum_{s=1}^{s=k} \left(\frac{\partial W}{\partial n_s}\right) dn_s \tag{9.9}$$

The problem with (9.9) is that the derivatives of W are difficult to formulate. It would be easier if we could take the logarithm of both sides and introduce $F = \ell n\, W$ as the function to be maximized:

$$F = \ell n \frac{n!}{n_1!\, n_2! \ldots} = \ell n\, n! - \sum_s \ell n\, n_s! \tag{9.10}$$

The distribution $\underline{n}$ which maximizes W will also be the one that maximizes F. Another help in the calculation is the introduction of Sterling's approximation,

$$\ell n\, m! = m\, \ell n\, m - m \tag{9.11}$$

which gets rid of factorials and leaves us with

$$F = n\, \ell n\, n - \sum_s n_s\, \ell n\, n_s \tag{9.12}$$

To get from (9.10) and (9.11) to (9.12) we had to make a cancellation by noticing

$$n = \sum_s n_s \tag{9.13}$$

The introduction of Sterling's approximation may raise eyebrows, as it has in the past. It does become exact in the limit of large numbers, so it is an "asymptotic" formula. But it is very bad for numbers as small as 10. For the kinds of numbers we encounter (Avagadro's number 6×10^{23}), the approximation is good for many more significant figures that could ever be measured at the macroscopic level. And for those special terms in the sum of (9.10) where n_s is down to the low levels where the approximation could fail, the values of those terms are negligible anyway. Sterling's approximation does limit us to "asymptotic" formulas, but those are the only kind ever offered or needed in statistical mechanics; there is no default in the use of Sterling's approximation.

So we can use (9.12) to find the most probable distribution. For any change $d\underline{n}$ we have

$$dF = \sum_s \left(\frac{\partial F}{\partial n_s}\right) dn_s \tag{9.14}$$

And when we have found the most probable $\underline{n}$ which maximizes F, then starting from there and deviating by $d\underline{n}$, we find

$$dF = 0 \tag{9.15}$$

We want to find $\underline{n}$ such that for changing $\underline{n}$ to $\underline{n} + d\underline{n}$ we get (9.15).

Our first problem is to formulate the derivative of F with respect to n_s. Let us start with the case of $s = 1$. Then we have

$$F = (n_1 + n_2 + \ldots)\ln(n_1 + n_2 + \ldots) - n_1 \ln n_1 - n_2 \ln n_2 - \ldots$$
$$\frac{\partial F}{\partial n_1} = 1 \ln n + \frac{n}{n}\, 1 - \ln n_1 + \frac{n_1}{n_1} = -\ln n_1 + \ln n = -\ln \frac{n_1}{n} \tag{9.16}$$

Certainly, then, we have, for any s:

$$\frac{\partial F}{\partial n_s} = -\,\ell n\left(\frac{n_s}{n}\right) \tag{9.17}$$

and we have the total change in F due to the contributions from all the dn_s as

$$-\,dF = \sum_s \left(\ell n\ \frac{n_s}{n}\right) dn_s = 0 \tag{9.18}$$

If the dn_s can have any arbitrary values whatsoever, then it follows that their coefficients in (9.18) must separately have the value zero:

$$\ell n\ (n_s/n) = 0 \tag{9.19}$$

and that implies, in the present context, that each n_s is equal to n, possible only if n and the n_s are all equal to infinity. That is not what we were after even though it does make good sense. Certainly the maximum value of W in (9.8) is reached when all variables have infinite value.

Yes, we did take it for granted that the number n was fixed as a constraint on the whole procedure. But we never did say that explicitly and incorporate it into the calculation, so we should not be surprised that the scheme gave a ridiculous answer to our question. We have to impose the restriction (constraint) that n is fixed. This can be expressed by

$$dn = \sum_s dn_s = 0 \tag{9.20}$$

which tells us that in our juggling of the n_s to find the maximizing distribution we can allow for all but one of the variations dn_s to be arbitrary but we must leave the last dn_s to be determined by (9.20). Thus we cannot assume that the coefficients of dn_s on the right side of (9.18) are all equal to zero. If we knew that one of them was zero, which would eliminate that term, then we could be assured that all the other dn_s were independently variable and hence all their coefficients must be zero, but we have no reason to believe that any one of those coefficients is zero.

Let us now systematically impose the constraint (9.20) while we try to "solve" (9.18) and find the maximizing distribution. There is a clever trick for doing this known as the "method of Lagrange multipliers," which turns out to be the backbone of statistical mechanics. The early steps seem stupid, and then we see how to turn them into a procedure that works.

First we reconsider (9.20) and notice the zero value. Then we multiply (9.20) by some number yet undetermined, a number α. Then we add to this (9.18) to get the perfectly valid but useless statement

$$- dF + \alpha\, dn = \sum_s \left(\ln \frac{n_s}{n} + \alpha \right) dn_s = 0 \qquad (9.21)$$

Are the coefficients of each dn_s on the right side separately equal to zero, which is the only kind of result that can give us a solution? The answer is negative because, again, the dn_s are not arbitrary in view of the connection (9.20). And the answer is even more negative when we recall that α can have any value (as yet undetermined). It would be very foolish here to argue that each of those coefficients of dn_s on the right side of (9.21) must be zero. But we can say this: Suppose we choose α to make the first coefficient equal to zero,

$$\ln \frac{n_1}{n} + \alpha = 0 \qquad (9.22)$$

Then that would eliminate that term with its dn_1 from (9.21) and leave only terms involving dn_2, dn_3, ... , all of which are entirely arbitrary, since dn_1 can always be selected in such a way as to satisfy the constraint (9.20). It follows that the coefficients of all the other dn_s (other than dn_1) in (9.21) must be zero, and we have the following set of equations

$$\ln(n_s/n) + \alpha = 0 \qquad (9.23)$$

for $s = 1, 2, 3, \ldots, k$. Thus we have our most probable distribution as

$$n_s = ne^{-\alpha} \qquad (9.24)$$

At first sight it is surprising that we could solve the maximization problem so easily. We were trying to evaluate k unknowns n_s without even having said what k is, and suddenly we have the answer in the simple form of (9.24). The achievement is truly remarkable, but a price was paid. We had to introduce a new unknown, namely α, and now we have to find α. When that has been done then we can say we have the solution of the maximization problem. We are grateful for the reduction from k unknowns to one unknown, but we still have to find that one unknown.

One could look at (9.24) and say that its message is that all the n_s have the same value, which is obviously n/k, and that completes the solution. Or one could proceed systematically as follows. The equation we invoke to be the one to solve for α is the original constraint that caused the need for α, namely,

$$n = \sum_s n_s = ne^{-\alpha} \sum_s 1 = ne^{-\alpha}k \qquad (9.25)$$

Cancelling the n we find

$$e^{-\alpha} = 1/k \qquad (9.26)$$

where k is the number of states available for the individual particle. Thus we have (9.24) as

$$n_s = n/k \qquad (9.27)$$

Notice that α played its remarkable role and then politely disappeared.

The most probable distribution of particles into states subject to the constraint that the total number of particles is fixed is given by (9.27). This implies that the distribution is uniform (same number of particles in every state). The system is "homogeneous." Actually that is the case for the spatial distribution of molecules in observed fluids at equilibrium without noticeable effects of external forces. Such fluids are uniform, for that is the overwhelmingly most probable distribution of the parts or particles.

The method described above is not restricted to fluids. We could just as well apply it to the distribution of people in a large hall, where many people are milling around somewhat randomly. If we ask them to simply move about without deliberate clustering they will approach a uniform distribution. This approach to uniformity is a familiar concept and it is that concept which causes many of our problems with understanding irreversibility. So let us digress for a moment to consider the situation in more detail.

Granted that uniformity is the final fate of a large group of random walkers, what can we say about the approach to that uniform, most probable distribution? It is tempting to think that this approach to "equilibrium" is necessary and reproducible and is determined by the future equilibrium to be melted into. But consider the case of a large number of people entering a large room through open doors. It could be that they are marching in straight lines that bring them to a configuration that continues to march around the periphery of the room without ever reaching equilibrium. Or it could be that they start that way but errors lead to wanderings and eventually we have our uniformity. That process is certainly not reproducible or predictable if we do not know a lot of details about the individual marchers. There is one and only one case where we can predict a reproducible evolution toward the future equilibrium. And that is the case where the people had already reached the uniform milling condition in the next room before the doors opened to the room we consider. They would mill in at a predictable rate in predictable directions and we could describe the diffusion leading to the new equilibrium uniformity. The prediction depends on the new "boundary conditions" (constraints) but not on the new equilibrium. It depends on the previous equilibrium (history of the sample).

Let us move on to the next problem. Suppose we have a large number of people in a large hall gambling with money. Let m_i be the wealth (number of dollars) of person i. Then there is a distribution of wealth $\underline{n} = (n_1, n_2, \ldots.)$ where n_m is the number of people who possess m dollars. The total wealth of the group is

$$M = \sum_i m_i = \sum_m n_m m \qquad (9.28)$$

and this, let us say, is fixed (a constraint). The gamblers win some and lose some, but as time passes, assuming the thing is fair and there is no "sink" which draws off money to a "house," then the distribution of wealth approaches an equilibrium which has the most probable distribution, and it stays there for a long time. Eventually the distribution may suddenly withdraw from equilibrium and do strange things, but then it will return for another long stay at equilibrium. The question for us here is: what is that equilibrium distribution of wealth?

What we have here is a question similar to the one described above requiring a most probable distribution in the sense of (9.18) with the constraint (9.20) but now we add a new constraint, namely the fixed level M of the total amount of money (9.28). Finding the most probable distribution is a matter of assuming (9.18) together with the constraint (9.20) and the new constraint (9.28) expressed as

$$dM = \sum_m m\, dn_m = 0 \tag{9.29}$$

If we multiply (9.29) by a number β and add that to (9.21), we get

$$-\, dF + \alpha\, dn + \beta\, dM = \sum_m [\ell n (n_m/n) + \alpha + \beta m]\, dn_m = 0 \tag{9.30}$$

There is no reason for the coefficients of dn_m on the right side of (9.30) to vanish unless we go out of our way to choose α and β to force the first two (any two) coefficients to vanish. That can be done, of course, and it would eliminate the first two terms from the right side of (9.30). Then we could say that the other dn_m are arbitrary, and hence the other coefficients must also be zero:

$$\ell n (n_m/n) + \alpha + \beta m = 0$$

or

$$n_m = n e^{-\alpha} e^{-\beta m} \tag{9.31}$$

Once again we witness the power of the simple method of Lagrange multipliers. The formula (9.31) reduces the problem of finding k unknowns n_m that maximize W under the restriction of two constraints (9.25) and (9.28) down to the problem of finding two parameters α and β which satisfy the constraints. To find or eliminate α we put (9.31) into (9.25) and get

$$n = ne^{-\alpha} \sum_m e^{-\beta m}$$

$$e^{-\alpha} = 1/q, \quad q = \sum_m e^{-\beta m} \tag{9.32}$$

This eliminates α by expressing it in terms of β. Incidently, the sum q is the power series for $1/(1 - x)$ where x is $\exp(-\beta)$. So we have

$$q = (1 - e^{-\beta})^{-1} \tag{9.33}$$

Putting (9.33) and (9.32) into (9.31) we find the most probable distribution of wealth as

$$n_m = n(1 - e^{-\beta})e^{-\beta m} \tag{9.34}$$

To find β we put this formula into the second constraint (9.28) and get

$$M = n(1 - e^{-\beta}) \sum_m me^{-\beta m} \tag{9.35}$$

as the equation to be solved for β. The sum on the right side is nearly the derivative of $q = 1/(1 - x)$:

$$\frac{d}{dx} \frac{1}{1 - x} = \frac{1}{(1 - x)^2}$$

and

$$\frac{d}{dx} \frac{1}{1 - x} = \sum_m mx^{m-1} = \frac{1}{x} \sum mx^m \tag{9.36}$$

Comparing these two formulas we have (9.35) as

$$M = n \frac{x}{1 - x}, \quad x = e^{-\beta} \tag{9.37}$$

Thus β turns out to be

$$x = e^{-\beta} = \frac{M}{M+n} \tag{9.38}$$

and the most probable distribution is

$$n_m = n \frac{n}{M+n} \left(\frac{M}{M+n}\right)^m \tag{9.39}$$

If we were interested in this problem of the maximum distribution of wealth we would notice some graphs and see that the most probable (gambling equilibrium) distribution is approached more closely in capitalistic economies than in communistic or third world economies. But of course in this context there is no reason to expect the method of statistical mechanics to offer important models. Our expectations grow when we return to the fluid composed of inanimate particles and the "wealth" to be distributed is something like energy. What we learned from the example above is that the most probable distribution is not always uniform. It was uniform when we had only the constraint (9.25). But the addition of (9.28) led to a result that was far from uniform.

Let us consider the foundation problem for statistical thermodynamics, which is similar in many ways to the distribution-of-wealth problem. Suppose we have a gas of n (fixed) particles and total energy E (all kinetic, and fixed). There are collisions in which kinetic energies change but at any one time there is no significant potential energy due to interactions and the total energy (kinetic) remains constant. In other words, we consider the classic case of the isolated system which is an ideal gas.

Thermodynamics always starts with such isolated systems and then develops derivations for moving to interacting systems. When we refer to the first law as a law about the "conservation" of energy, for example, we are referring to that primordial isolated system, for energy is certainly not conserved in the interacting systems of interest. It is conserved on a larger scale in the sense that the energy one system loses is equal to the energy another system gains. But before reaching for that we consider the simple isolated system with fixed energy. Let the particles collide and exchange the energy wealth until equilibrium has been reached where all exchanges of one kind are matched by exchanges of the opposite kind. The distribution of energy over the molecules at that point should be the most probable distribution for fixed n and E.

Our first problem is to formulate the energy constraint. This is similar to the money constraint of (9.28),

$$E = \sum_i \epsilon_{s_i} = \sum_s n_s \epsilon_s \qquad (9.40)$$

where ϵ_s is the energy of one molecule in states. Thus the total energy E can be regarded as a sum over particles i of their energies or a sum over particle states s of the energies ϵ_s multiplied by the number n_s of particles in state s. When we consider the various variations dn_s of the distribution in our search for the distribution that maximizes W we may include only variations that satisfy

$$dE = \sum_s \epsilon_s dn_s = 0 \qquad (9.42)$$

So now we look for the distribution $\underline{n}$ that maximizes W subject to the constraints (9.25) and (9.40). This means we look for the $\underline{n}$ that gives (9.18) for all $d\underline{n}$ that satisfy (9.20) and (9.42). Using the method of Lagrange multipliers, we have the analog of (9.30) as

$$- dF + \alpha\, dn + \beta\, dE = \sum_s [\ln(n_s/n) + \alpha + \beta\epsilon_s]\, dn_s = 0 \qquad (9.43)$$

Repeating the argument leading to (9.31) we point out that although it is not necessary for each of the coefficients of the dn_s on the right side of (9.43) to have zero values, it is, nevertheless true that if any two of the coefficients were zero then it would be necessary for all the rest to be zero. Since we have two "adjustable parameters" α and β at out disposal, we can choose them to force two coefficients to be zero and acquire the remarkable equations

$$\ln(n_s/n) = - \alpha - \beta\epsilon_s$$

$$n_s = ne^{-\alpha-\beta\epsilon_s} \tag{9.44}$$

Thus the problem of finding the most probable distribution of gas particles (molecules) over states in the overwhelmingly most probable distribution which prevails at equilibrium and nearly prevails for all gas situations is solved by a formula with two parameters α and β. There have been experiments to test this result but they were foolish, in retrospect. It can be shown from statistics alone that there is no way for (9.44) to be wrong for an ideal gas at equilibrium, even if not isolated. The strangest virtue of probabilistic methods, properly handled, is that they can tell you when they are so sure of themselves that experimental confirmation is unnecessary. And it is a rare situation where they can say that. Perhaps, only in the case of statistical mechanics. The point here is that (9.44) does provide the equilibrium distribution of molecules over states without the slightest uncertainty.

We still have the problem of finding the Lagrange multipliers α and β. To find or eliminate α we do the same thing we did above at (9.32) and we find

$$e^{-\alpha} = 1/q \qquad q = \sum_s e^{-\beta\epsilon_s} \tag{9.45}$$

and we get

$$n_s = \frac{n}{q} e^{-\beta\epsilon_s} \tag{9.46}$$

Here we cannot formulate the sum q as we did in (9.33) because ϵ_s is not generally proportional to an integer. It is in the case of an "ideal gas" of oscillators and that is why Planck could finish his statistical mechanics for black-body radiation and Einstein could develop his model for the heat capacity of crystals. But more generally we cannot evaluate q without more difficulty, if at all. So we find no difficulty in eliminating α but we expect difficulty in finding β.

To find β we need the second constraint (9.40) in the form

$$E = \sum_s n_s\epsilon_s = \frac{n}{q} \sum_s \epsilon_s e^{-\beta\epsilon_s} \tag{9.47}$$

If we make q explicit in terms of the unknown β and see (9.47) as an algebra problem to be solved for β with everything else known, we see an algebra problem that makes the invention of logarithms look like a Sunday School picnic. There is no hope of solving that equation. It was the genius of Boltzmann, generalized many ways by Gibbs, that turned the problem around. Here we see n and E, along with the ϵ_s, as the given independent variables, and we see β as a dependent variable to be found in the sense of finding its formula in terms of n, E, and the ϵ_s. But actually E itself is not measurable or controllable. Why should E be an independent variable? Let us abandon the search for β as a function of E and keep β itself as an independent variable, retiring E to the role of a dependent variable. Then (9.47) is itself already the formula we need for E in terms of β.

This switch in the roles of E and β is fine up to a point. After all, a Lagrange multiplier like β is just a helping function, an intermediate in the calculation. Elevating that servant to the status of master requires a lot of luck.

As it happens, the luck is there. The parameter β turns out to be temperature, which is much more measurable and controllable than E. And what we learn out of that is that our old friend temperature is actually a Lagrange multiplier, or a parameter for a most probable distribution at equilibrium. In retrospect we admit that the temperature had to be something like that, for it does not occur in the Newtonian model for the molecules. We see that temperature is intimately tied to equilibrium and its important role in fluid dynamics suggests important ties between fluid dynamics and equilibrium. Temperature is bound to be close to the heart of our larger problem here, that of explaining determinism and irreversibility in fluid behavior.

At this point, if we were interested in statistical thermodynamics for its own sake, we would go on to find the formula for the internal energy U of thermodynamics in terms of temperature and other variables, the formula for pressure, the formulas for the enthalpy and entropy and the various free energies. We cannot indulge in that here, but we can report that the formula for β in terms of the Kelvin temperature T is

$$\beta = \frac{1}{kT} \tag{9.48}$$

and the formula for entropy on Boltzmann's tombstone is well founded:

$$S = k \ln W \tag{9.49}$$

The proportionality coefficient k in (9.48) is now a well-known and "universal" constant known as Boltzmann's constant (the gas constant R divided by Avagadro's number). The number W in (9.49) is the value of W for the most probable distribution, even away from equilibrium (which contributes to the mystery of entropy).

9.2 Generalization to Systems of Interacting Particles

The method described above is flawless in a sense and does lead to a successful prediction of equilibrium properties, equations of state, etc., for ideal gases. It even works for some models of crystals. The crucial assumption is that the energy can be expressed as a sum of separated contributions, as in (9.40). This is the case for an ideal gas, where the molecules have the separated energies, and for other collections such as the perfect crystal, which can be regarded as an ideal gas of oscillators, ideal in the sense of having a formula like (9.40). But as soon as we lose (9.40), we lose the whole ball game. We lose the method of Lagrange multipliers, and no one has ever figured out how to do statistical mechanics without that. The two methods are essentially the same thing in our context.

Getting beyond the severe restriction of (9.40) required the genius of J.W. Gibbs (1902). Having put together thermodynamics in its final form 25 years earlier, he then put together statistical mechanics in its final form with no restriction to (9.40), even though he depended on (9.40) to develop the method, as he had to, for there was no other way. What he had noticed was that the average values of molecular properties for ideal gases were quite different from the values associated with individual molecules (there was a wide spread in the distribution), but the variations decreased with increasing complexity within the molecule. In other words, the energies of monatomic gas atoms were spread all over the map, with the average meaning nothing in the prediction of energy for one molecule. But when the molecule contains many atoms and has many internal "degrees of freedom," the deviation from average for randomly selected molecules was a lot less. In fact, if you go to molecules that have Avagadro's number of atoms, they all have the average values of any property. A collection of them would be very dull, for they would all have the same property values, very unlike a gas of Ne atoms, or even NH_3 molecules.

Suppose you want to describe a beaker of water. The thing to do, according to Gibbs, is to imagine a collection of trillions of beakers of water connected in the sense that they are touching and can exchange energy but not connected "forcefully" in the sense that one could exert a force on the other. This is an "ideal gas" of beakers of water, very hypothetical, admittedly, but nevertheless a collection ("ensemble") that would be like the collection considered above, except for the fact that the individual molecules above are now beakers of water, i.e., very large molecules.

We could look for the most probable distribution of beakers of water over their states in exactly the same manner as above and end up with the same formulas, but to remind ourselves that here the "molecule" is a whole beaker of water, and its state is really a superstate compared to the state s of a small molecule, let us refer to the state S of a beaker, and the energy E_S of a beaker in state S, and the distribution $\underline{N} = (N_1, N_2, \ldots.)$ of beakers of water into their states (N_S is the number of beakers in state S). The total number N of beakers of water has no physical significance; in fact it can be infinite causing asymptotic formulas to become strictly exact.

The problem is to find the distribution $\underline{N}$ that maximizes the super W,

$$\mathcal{W} = \frac{N!}{N_1!\,N_2!\,\ldots} \tag{9.50}$$

for fixed values of the total number N and the total energy $\mathcal{E}$ of the ensemble. Here we have the analogs of (9.25) and (9.40) as

$$N = \sum_S N_S \qquad \mathcal{E} = \sum_S N_S E_S \tag{9.51}$$

Applying the method of Lagrange multipliers in exactly the manner as above, we find the most probable distribution as

$$N_S = Ne^{-\alpha-\beta\varepsilon_S} \tag{9.52}$$

This most probable distribution for an isolated ensemble of hypothetical supermolecules (beakers of water) is known as the Canonical Ensemble distribution. The "ensemble" and "canonical" terminology could be applied just as well to the ideal gas collection of small molecules described by (9.44), but there is a tendency to reserve the terminology for Gibbs' hypothetical collections of hypothetical systems.

Up to this point the mathematics of the Gibbs model and the Boltzmann model are identical. At this point, however, the Gibbs model takes a new turn. Rather than follow the values of N_s, we go for the ratios

$$P_S = N_S/N = e^{-\alpha-\beta E_s} \tag{9.53}$$

and then forget N_s and N, which referred to a hypothetical ensemble of no direct physical interest. The ratio (9.53) can be interpreted as the probability that any one beaker of water, picked at random, is in state S. This beaker picked at random is the one real beaker of water we are trying to describe, and what we now have are the probabilities of its states if it is in "thermal equilibrium" with a "reservoir." The abstract ensemble of statistical mechanics becomes the thermal reservoir of thermodynamics. If we now want to predict the properties of our lonely beaker of water we simply predict them as averages using the probabilities (9.53). For example, the internal energy U of thermodynamics is the average of E over the ensemble:

$$U = \sum_S P_S E_S \tag{9.54}$$

What makes this method work is that the average values become the <u>only</u> values (to many significant figures) for molecules containing Avagadro's number of atoms. This was not the case for small molecules. If we predicted that the energy of a small molecule in an ideal gas as the <u>average</u> energy, we would be making a huge error. But for large enough molecules the error is as small as you wish. And if you want to predict the entropy S for the one real beaker of water in the Gibbs ensemble, just calculate the entropy of the whole ensemble (the whole reservoir):

$$\mathcal{S} = k \ln \mathcal{W} \tag{9.55}$$

and divide by N. Using Sterling's approximation (9.11) we can derive

$$S = -\sum_{S} P_S \ln P_S \tag{9.56}$$

That formula, incidentally, is the one taken from statistical mechanics to form Information Theory, which deals with applications of the method of statistical mechanics to problems that are not thermodynamic.

It we were aimed at the applications of statistical mechanics, we would finish what was started above and try to find α and β in (9.53). We could find α and it would turn out to be the analog of (9.45) with q replaced by

$$Q = \sum_{S} e^{-\beta E_S} \tag{9.57}$$

This is the "canonical partition function" where q was merely the "partition function." (We could refer to Q as the superpartition function.) The attempt to find β would follow the same route as in the Boltzmann ideal-gas model and we would end up with (9.48) expressing β in terms of the more familiar temperature T (Kelvin). One could then say that the "thermal reservoir" of Gibbs is a hypothetical collection of systems in thermal contact which all have the same temperature. The "Lagrange multiplier" interpretation of temperature is elevated to the "common property" interpretation.

The method of Gibbs does put statistical mechanics into its final form at the level of general theoretical development. It covers all equilibrium situations and it can be modified to make it more convenient toward new choices of independent variables, new considerations of external forces, etc. It seems to work a miracle. But there is one hitch. In order to be able to apply the method, we have to be able to evaluate the sum in (9.57), to find the formula for Q in terms of thermodynamic variables used to describe the system. The analogous summation in (9.45) over states of simple molecules can be carried out in many cases leading to an extensive application of statistical mechanics to ideal gases. But when the states are the states of beakers of water and the energy formulas are at that level, the possibility of carrying out the summation is almost hopeless. It has been done for a few special cases. For future applications, it is easy to identify the basic problem: Find Q.

It is interesting to notice that Einstein was especially interested in the generalization of the Boltzmann statistical mechanics and in the use of statistical mechanics to illustrate kinetic theory, quantum mechanics, etc. He is one of the few people at the turn of the century who really understood the method and its power. In fact, he came up with a generalization equivalent to the one developed by Gibbs. But then he modestly stepped back and withdrew his approach. On the other hand, we have some caustic criticism of the "ensembles" and "reservoirs" of Gibbs by Khinchin (1949) who said that Boltzmann was "timid" and Gibbs was "naive," unaware of the methods then available in probability theory. Why give names to ensembles when everything in probability work refers to some hypothetical ensemble, of course? All that was ever needed is an awareness of the powerful "central limit theorem." But Khinchin never displays an awareness of the difference between equilibrium and nonequilibrium, and he also misses quite a few other things that account for our monumental debt to Boltzmann and Gibbs.

We notice that throughout this section we have ignored the time evolution of particles, molecules, beakers of water, whatever. We were looking for most probable distributions, which are certainly the distributions that prevail at equilibrium. Our demands on mechanics were quite limited, but our demands on probability ideas were very strong. We were totally preoccupied with equilibrium and learned nothing about withdrawals from or approaches to equilibrium or anything else about the evolution of systems as they change in time. We learned the whole thing about equilibrium, except for a vast collection of knowledge about details of applications, but that was all aimed directly at equilibrium, which is the "home ground" for statistical mechanics (the method of equilibrium).

In Section 9.4 we go to the other extreme. We may borrow the idea of a distribution from these past sections, but our question is not what the distribution is at any one time. The question is: how does the distribution, whatever it was, change in time if the molecules follow Newtonian prescriptions for change?

But before we get into the mechanics of Section 9.4, let us consider briefly what happens to our formulas above if the states are labeled by variables s or S which are not discrete but are continuous. That is our task in Section 9.3.

9.3 Generalization to a Continuum of States

The most probable distribution of molecules over states in an isolated ideal gas is given by (9.46). Acknowledging the switch (9.48) from statistical temperature β to Kelvin temperature T, we have

$$n_s = \frac{n}{q} e^{-\epsilon_s/kT} \tag{9.58}$$

$$q = \sum_s e^{-\epsilon_s/kT} \tag{9.59}$$

Suppose the states labeled above by discrete values of s are actually points in a continuum. This would be the case if our particle were a particle with position x (one dimension) and velocity v. Then s is (x,v) where x and v have a continuum of values. Maybe the sum (9.59) can simply be replaced by an integral,

$$q = \int dx \int dv\, e^{-\varepsilon(x,v)/kT} \tag{9.60}$$

but how do we formulate n_s in (9.58)? This seems like a trivial question, but it was answered poorly before 1900 so many times we have to be sensitive about it, and it required the advent of quantum mechanics to make us sensitive about the question.

First of all, we have to notice that we cannot just willy-nilly replace the sum over s in (9.59) by an integral over x and v. We could replace the sum over s in (9.45) by a sum over <u>cells</u> in the continuum state-space of (x,v) if we include a factor G(x,v) which is the number of states in each cell centered around x and v. Better yet, let us express G as the "volume" (here area) dxdv of a small cell multiplied by g(x,v) which is the density of states in the state-space:

$$q = \sum_{\text{cells}} g\,dxdv\, e^{-\varepsilon/kT} \tag{9.61}$$

Then we have the integral form of (9.59) as

$$q = g \int dx \int dv\, e^{-\varepsilon/kT} \tag{9.62}$$

This assumes that g is a constant over the space of x and v. That assumption seems reasonable and it was confirmed eventually by quantum mechanics with its "uncertainty principle" which says that the area (uncertainty) of one state in our state-space is:

$$\Delta x\, \Delta p = h \tag{9.63}$$

where the momentum p is mv. So we have the "area" of one state as

$$\Delta x\, \Delta v = h/m \tag{9.64}$$

and the number of states in an area dxdv as

$$q = \frac{dxdv}{\Delta x \Delta v} = (m/h) \tag{9.65}$$

which is constant (uniform in x and v). Since we are not interested here in quantum mechanics we will not use (9.65) but we will use g in (9.61) and (9.62) as a constant which can be ignored (call it 1). So we join our classical colleagues in assuming (9.60). But there is still the problem converting (9.58) to continuum form.

We could say that the number of particles in state s, with positions between x and x + dx, velocities between v and v + dv, is n_s expressed as a density f_s multiplied by dxdv. Then the density f_s is shown by (9.58) to be proportional to the exponential (exp - ε/kT),

$$f(x,v) = Ae^{-\varepsilon/kT} \tag{9.66}$$

leaving A to be found as a "normalization constant." This is a dangerous assumption. In some contexts similar to the present one, A turns out to be proportional to v, or something like that. One has to be careful when translating discrete formulas to continuum formulas, or vice versa. But in the problems of statistical mechanics we always luck out, and the supporting proofs are available. So we do have (9.66), and, in fact, we have a simpler formula. For the systems we consider, there is no external force and the distribution is uniform in space (same at all x), so we can present (9.66) as

$$f(v) = Ae^{-\varepsilon/kT} \tag{9.67}$$

The energy ε is purely kinetic,

$$\varepsilon = \frac{1}{2} mv^2 \tag{9.68}$$

Thus we have the distribution of molecules (the density in v-space) over velocities as

$$f(v) = Ae^{-(m/2kT)v^2} \tag{9.69}$$

To find the value of proportionality coefficient (normalization constant) A, we notice that the sum over all velocities v of the number fdv of molecules with velocities between v and v + dv is the total number of molecules n:

$$n = \int_{-\infty}^{\infty} f(v)\, dv = A \int_{-\infty}^{\infty} e^{-(m/2kT)v^2}\, dv \tag{9.70}$$

Using the integral formula

$$\int_{-\infty}^{\infty} e^{-av^2}\, dv = \sqrt{\frac{\pi}{a}} \tag{9.71}$$

we have (9.70) as

$$n = A\left(\frac{2\pi kT}{m}\right)^{1/2} ,$$

$$A = n\left(\frac{m}{2\pi kT}\right)^{1/2} \tag{9.72}$$

Thus we have the one-dimensional "distribution function" f as

$$f = n\left(\frac{m}{2\pi kT}\right)^{1/2} e^{-(m/2kT)v^2} \tag{9.73}$$

The same derivation can be developed in two and three dimensions to find the normalized most probable distribution functions. In three dimensions we have the famous Maxwellian formula

$$f = n\left(\frac{m}{2\pi kT}\right)^{3/2} e^{-(m/2kT)v^2} \tag{9.74}$$

where

$$v^2 = v_x^2 + v_y^2 + v_z^2 \tag{9.75}$$

The primary formula (9.74) is presented in a variety of contexts with slightly different meanings. In some contexts, n is not the total number of particles but the number per unit volume. In other contexts, n is absent, and the rest of the formula (9.74) is the probability that a molecule picked at random has the velocity v. One simply has to be careful to notice the context.

When we move to the Gibbsian view of super molecules (the whole system) distributed over superstates, it is customary to alter the notation in certain ways. The state space of S is the 6n-dimensional space of

$$S = (\underline{s}_1, \underline{s}_2, \underline{s}_3, \ldots) \qquad \underline{s}_i = (\underline{r}_i, \underline{v}_i)$$

$$\underline{r}_i = (x_i, y_i, z_i) \qquad \underline{v}_i = (v_{xi}, v_{yi}, v_{zi}) \tag{9.76}$$

We could also use the notation

$$S = \underline{s}^n = (\underline{r}^n, \underline{v}^n)$$

$$\underline{r}^n = (\underline{r}_1, \underline{r}_2, \underline{r}_3, \ldots) \qquad \underline{v}^n = (\underline{v}_1, \underline{v}_2, \underline{v}_3, \ldots.) \tag{9.77}$$

The probability of the superstate S at equilibrium, given by (9.53), for discrete states, is replaced by a probability density P(S) defined by saying that P(S)dS is the probability that the state S of the whole system is "between" S and S + dS, i.e., within a cell of "volume" dS containing S in the superstate space. This is

$$P_S dS = P(\underline{s}^n)d\underline{s}^n \tag{9.78}$$

where the differential "volume" element $d\underline{s}^n$ is

$$d\underline{s}^n = d\underline{s}_1 d\underline{s}_2 \ldots d\underline{s}_n$$

$$d\underline{s}_i = d\underline{r}_i d\underline{v}_i$$

$$d\underline{r}_i = dx_i dy_i dz_i \qquad d\underline{v}_i = dv_{xi} dv_{yi} dv_{zi} \tag{9.79}$$

What we are trying to do is to present the probability P_s in (9.53) as a thing we can talk about when the states S form a continuum. Maybe what we should have said was simply that P_s is proportional to $\exp(-\beta E_s)$ and the proportionality coefficient will be found when needed by normalizing. As for β, it is, of course, as always, (1/kT). On the other hand, we will need some of the notation introduced above in later discussions.

What we have here is that the probability (density) of a super distribution $\underline{N}$ over the states S of N systems (super molecules) has the same form (proportional to the negative of the exponential of energy divided by kT) as encountered above for the ideal gas. Thus we have

$$P(S) = Ae^{-E_s/kT} \tag{9.80}$$

The normalization constant A is not important here and we will not try to evaluate it.

Moving from discrete distributions to continuous distributions is more of a nuisance than we might have expected. But it could be a lot worse. It turns out that every move we want to make can be made somewhat naively because of certain features of Newtonian mechanics. Workers who have sought to follow other kinds of distributions (bubble sizes in foams, crystal sizes in melts) have been misled by blindly following the transitions from discrete to continuous in statistical mechanics. The main problem for students of statistical mechanics is the expanding notation. In our own case, we have the formulas we need and will leave this problem at this point.

9.4 The Liouville Theorem

Suppose we know the distribution of one or many particles over the appropriate states at the initial time. How does that distribution change in time? We admit no stochastic flukes that violate Newton's law but insist that the evolution is strictly Newtonian.

First let us consider a collection of particles moving in one dimension without interactions. Let us say that f(x,v,t) is the number density in the state-space of (x,v), sometimes called the "distribution function." In other words, f dxdv is the number of particles with positions in the range dx around x and with velocities in the range dv around v at time t. If we follow that element as it moves in the state-space of (x,v), it must continue to contain the same number of particles. This is the conservation of matter expressed by (8.19), and by the reasoning used there we arrive at the inevitable "equation of continuity,"

$$\frac{\partial f}{\partial t} + \frac{\partial}{\partial x}(vf) = 0 \tag{9.81}$$

It does not matter that we are here in a more complex space. The equation of continuity (8.20) holds in all spaces for all fluids so long as there are no creation or destruction processes involved, so long as the moving fluid element retains its innards and receives no new material. What we have here that we did not have in the context of (8.20) is the remarkable equation

$$\frac{\partial}{\partial x} v = 0 \tag{9.82}$$

which leaves (9.81) in the simpler form,

$$\frac{\partial f}{\partial t} + v \frac{\partial}{\partial x} f = 0 \tag{9.83}$$

known as Liouville's equation.

In the context of ordinary fluid dynamics, (9.81) holds always but (9.82) and (9.83) hold only in the special case where the fluid is "incompressible." The distinction between compressible and incompressible fluids is a sensitive issue in fluid dynamics. One could say that it is the difference between the nightmare of aeronautical engineers, who deal with compressible gases, and the situation of chemical engineers, who enjoy incompressible liquids. In any case, (9.83) is much simpler in our context than (9.81), and we have the preferred equation for a reason that has nothing to do with the compressibility of real fluids. We have (9.83) because we are working in the mechanical state-space of x and v where v is an <u>independent</u> variable that cannot be a function of x. Thus we can visualize f as the density of a fluid of some kind (probability fluid) and we can assume that the fluid is incompressible. That is Liouville's theorem.

The solution of a first-order, linear partial differential equation like (9.83) can be developed in various ways, some of which will be important for us in more complex contexts. First of all there is the method of the mathematical theory (Webster, 1933, Chapter 1) which goes as follows. Notice the coefficients 1 and v in front of the derivatives in (9.83). Divide each by the lower differential of its term and equate the results:

$$\frac{dt}{1} = \frac{dx}{v} \tag{9.84}$$

and treat that as the equation for the "trajectories." Here we have

$$dx/dt = v \qquad x = x_o + vt \tag{9.85}$$

Then any function of the constant $x - vt$ is a solution of (9.83). The initial value x_0 is a constant along the "trajectories" defined by (9.85).

It is easy to confirm the recipe given above, but it takes some thought to choose the "any function." Consider any function $g(y)$, where y is $x - vt$. Then we have

$$\frac{\partial g}{\partial t} = \frac{\partial g}{\partial y}(-v), \quad \frac{\partial g}{\partial x} = \frac{\partial g}{\partial y},$$

$$\frac{\partial g}{\partial t} + v\,\frac{\partial g}{\partial y} = \frac{\partial g}{\partial y}\,(-v+v) = 0 \qquad (9.86)$$

This confirms the recipe but leaves us wondering what g is. If we are clever we can figure that out. But even if we are not we can consider another approach which offers g from the start.

Suppose we know f at the initial time, call it F(x,v):

$$f(x,v,0) = F(x,v) \qquad (9.87)$$

Notice that v is just a floating constant parameter so far as (9.83) is concerned. We could have suppressed it from notation in (9.87). Why not choose F(x) as our g. Then we have g as F(x - vt). In other words, we have the final solution of (9.83) as

$$f(x,v,t) = F(x-vt,v) = f(x-vt,v,0) \qquad (9.88)$$

This does solve the problem, and hopefully our description was not too obscure.

And there is another way to solve (9.83) which is closer to applications in more complex systems but which does demand more of our patience toward methods that are, well, maybe correct, but far from the rigorous elements of calculus. We refer here to the language and notation of "operators" according to which we see (9.83) as

$$\frac{\partial f}{\partial t} + Lf = 0 \qquad (9.89)$$

where the Liouville operator is

$$L = v\,\frac{\partial}{\partial x} \qquad (9.90)$$

If L were a constant multiplying operator instead of a differentiating operator, we would say that the solution of (9.89) is

$$f(t) = f(0)e^{-Lt} \qquad (9.91)$$

Let us keep the same form but see L as the operator on the right side of (9.90):

$$f(x,v,t) = e^{-tv\frac{\partial}{\partial x}} f(x,v,0) \tag{9.92}$$

At first sight, the operator on the right side of (9.92) is meaningless. Whoever heard of the exponential of differentiation? But if we take the admittedly artificial symbolism literally, and represent the exponential by its Taylor series, we have

$$e^{-tv\frac{\partial}{\partial x}} g(x) = \left[1 - tv\frac{\partial}{\partial x} + \frac{(tv)^2}{2}\frac{\partial^2}{\partial x^2} + \ldots.\right] g(x) = g(x - vt) \tag{9.93}$$

Thus the exponential operator can be regarded as a "displacement" operator, displacing x back in time to x - vt. In fact, we can define the exponential operator as the displacement operator even for contexts where all the derivatives in (9.93) do not exist. It has also been given the fancy name of "time development operator." What it really is, is the go-back-along-the-trajectory-for-a-time-t operator. Thus the solution (9.92) is a fancy form of (9.88).

If we move to three dimensions, the problem becomes more interesting. It is the problem of the rarefied gas (Biolsi, 1974). We start with the always-valid continuity equation

$$\frac{\partial}{\partial t} f + \frac{\partial}{\partial \underline{r}} \cdot \underline{v} f = 0 \tag{9.94}$$

and then notice the incompressibility of the "fluid" (or the fact that $\underline{v}$ here does not depend on $\underline{r}$)

$$\frac{\partial}{\partial \underline{r}} \cdot \underline{v} = 0 \tag{9.95}$$

and find the Liouville equation. The Liouville operator L in (9.89) has become

$$L = \underline{v} \cdot \frac{\partial}{\partial \underline{r}} \tag{9.96}$$

and the solution of (9.94) is

$$f(\underline{r},\underline{v},t) = e^{-tL} f(\underline{r},\underline{v},0) = f(\underline{r} - \underline{v}t,\underline{v},0) \tag{9.97}$$

This formula can be used to follow the evolution of a rarefied gas from any initial condition. But it offers only the Newtonian mechanics for such a project. It tells us nothing about how to select an appropriate initial distribution.

If we move to a gas of n interacting particles, the problem becomes even more interesting. In fact it reaches the level of model development which is still controversial. Here we must ask the reader to visualize (by analogy) a hypothetical probability fluid with density $P(S,t)$ in the 6-n dimensional space of $S = \underline{s}^n$ which is the density in that space of points representing systems in the Gibbsian ensemble, or better yet, the probability density for the superstates of one supersystem (e.g., beaker of water), and look for the rate of change of P.

First we may say that there must be the usual equation of continuity

$$\frac{\partial P}{\partial t} + \frac{\partial}{\partial S}(PW) = 0 \tag{9.98}$$

where S is the superstate position and W is the velocity of such a superstate along a trajectory in the superstate space. To be more precise, we do have S as $\underline{s}^n$ and we have W as the Newtonian rate of change of S with time,

$$\begin{aligned} W &= \frac{d}{dt}(\underline{r}_1, \underline{r}_2, \underline{r}_3, \ldots, \underline{v}_1, \underline{v}_2, \underline{v}_3, \ldots) \\ &= (\underline{v}_1, \underline{v}_2, \underline{v}_3, \ldots, \underline{a}_1, \underline{a}_2, \underline{a}_3, \ldots) \end{aligned} \tag{9.99}$$

where the accelerations $\underline{a}_i$ are forces $\underline{F}_i$ divided by the mass m. The forces can often be expressed as $\underline{r}_i$ derivatives of a potential energy. We assume here that the forces (and potential energy) may depend on positions $\underline{r}_i$ but not on velocities $\underline{v}_i$. Taking into account the incompressibility of the fluid,

$$\frac{\partial}{\partial S} W = \sum_i \left(\frac{\partial}{\partial \underline{r}_i} \cdot \underline{v}_i + \frac{\partial}{\partial \underline{v}_i} \cdot \underline{a}_i \right) = 0 \tag{9.100}$$

we can replace (9.98) by the simpler Liouville equation,

$$\frac{\partial}{\partial t} P + LP = 0 \tag{9.101}$$

where the n-particle Liouville operator is

$$L = \sum_{i=1}^{n} \left(\underline{v}_i \cdot \frac{\partial}{\partial \underline{r}_i} + \underline{a}_i \cdot \frac{\partial}{\partial \underline{v}_i} \right) \tag{9.102}$$

The solution is, simply,

$$P(\underline{s}^n, t) = e^{-tL} P(\underline{s}^n, 0) \tag{9.103}$$

In view of the enormity of what we just achieved by notational tricks, we must pause to review some of the steps. We just solved a partial differential equation in 6n+1 variables, where n is Avagadro's number (10^{23}) and the solution involved solving Newton's equations for all those interacting particles to find trajectories. First of all, the incompressibility of the fluid with density P expressed in (9.100) followed from the fact that the $\underline{v}_i$ do not depend on the $\underline{r}_i$ and the $\underline{a}_i$ do not depend on the $\underline{v}_i$, even though they do depend on the $\underline{r}_i$. That leaves the equation (9.98) in the simpler form of the Liouville equation (9.101) which is a linear combination of derivatives like (9.83) except for being more complicated. The analog of (9.84) is

$$\frac{dt}{1} = \frac{d\underline{r}_1}{\underline{v}_1} = \frac{d\underline{r}_2}{\underline{v}_2} = \ldots = \frac{d\underline{v}_1}{\underline{a}_1} = \frac{d\underline{v}_2}{\underline{a}_2} = \ldots \tag{9.104}$$

which offers a set of equations

$$\frac{d}{dt} \underline{r}_i = \underline{v}_i \qquad \frac{d}{dt} \underline{v}_i = \underline{a}_i = \frac{1}{m} \underline{F}_i \tag{9.105}$$

for the trajectories of coordinates, and collectively, trajectories of states in the superstate space. We notice that (9.105) is the collection of Newton equations for the n particles. The trajectories referred to here are the Newtonian trajectories. Thus the Liouville operator in (9.103) is simply telling us to take the point $\underline{s}^n$ backwards along the n-particle trajectory for a time t to where that point was initially and assign to P the initial value of P at that initial point. Since the fluid there then evolved to become the fluid at $\underline{s}^n$ at time t, and without a change in the volume of the fluid element, the density P there then must be the same as the density at $\underline{s}^n$ now at time t. This is a very simple idea nearly smothered by notational difficulties.

The incompressibility of the probability fluid (9.100) is the celebrated Liouville theorem and the simplified equation (9.101) is the celebrated Liouville equation (Tolman, 1938, p.48). The trajectory solution is the fundamental starting point for the BBGKY theory (Section 4.4 above). This is all very impressive. And yet it is not clear that any of the derivations of this section have anything to do with statistical mechanics. Furthermore, no formula given here is deterministic at the real fluid level and no collection of these formulas is capable of carving out from the set of all possible Newtonian evolutions that subset which actually occurs in nature. The Liouville equation, by itself, is innocuous in the sense of offering no new obstacles to our quest for an understanding of irreversibility except for the intimidation of students, but it offers no help either. It is merely the Newtonian equations of motion for n particles expressed in a collective way. We hope to find a way to use it, but that will require a context we do not yet have.

There is one way to use the Liouville equation to try to learn something about equilibrium. This is a common technique in the fluid dynamical schemes for real fluids, so we will try it here.

Given the equation of change (the Liouville equation) [Eq.(9.101)], what do we find if we set the time derivative equal to zero? This gives the equilibrium Liouville equation

$$LP = 0 \tag{9.106}$$

What about the solutions of that equation? One thing we can say right off is that $LM = 0$ is solved by any function M(S) which is a "constant of the motion" or constant in time along a trajectory. So if P is any function of M, then it satisfies (9.106). In other words, P can vary with the $\underline{s}_i$, but only in a way that "goes through" M. It depends first on M and then through M on the $\underline{s}_i$. This is the case for (9.80), where M is the energy E, which varies with the $\underline{s}_i$ (is a function of those) but does not vary in time. It is very tempting to try to derive (9.80) from (9.106) and avoid all of what we called the method of statistical mechanics in the previous sections, and attempts to do this have been developed. The problem is in the selection of M. How could one know to use the energy? And then how to get the energy into the exponent of (9.80)? The arguments are pretty good. But how to know that energy only is the M that counts? According to (9.106), P can be any function of all possible M functions that are constants of the motion. Carried to the extreme, we could hope for as many such constants as there are coordinates and then the idea has been diluted far beyond any kind of usefulness in fluid dynamics. We always seem to need guidance in some form from the equilibrium statistical mechanics. Without that, the Liouville equation would have nothing to say. In our opinion, (9.106) has been highly overrated.

9.5 Joining Statistics and Mechanics: The One-Particle Approximation

To begin this discussion we have to make a choice which is worth noticing. Should we deal with the probability density P over the states of one particle, trying to find out where the particle is at later times, or should we follow many noninteracting particles with a density f in the state-space of one particle. In a way there is no difference. But conceptually it is nicer to deal with f and that is our choice. When we deal with larger systems composed of interacting particles, we may choose P because it is easier to normalize.

The evolution of f is given by the Liouville equation (9.83) leading to (9.88) or its equivalent (9.92). If we start with that and ask for fluid equations of equilibrium, the only answer we can get is that f must be uniform in x. That is implied explicitly by (9.83) if the time derivative of f is zero. This is the correct conclusion in our context, in which external forces are being ignored, but it only gives us the spatial uniformity of equilibrium. It cannot display the effects of an energy constraint or any other constraint that would shape the velocity distribution. So let us admit that we are going to learn nothing important about equations of equilibrium from the Liouville equation. That presents no problem, for we have a complete theory about such equations based on the method of statistical mechanics.

Consider next the fundamental equations of change [Eqs.(8.2) - (8.5)], which have no dependence on statistical mechanics or equilibrium. Since we already have those equations from applications of basic laws to the fluid continuum, one could argue that there is no need for further information from the Liouville equation. On the other hand, it is surprising that the fluid continuum concept can work as well as it does because it ignores the uncorrelated flights of neighboring molecules through the vacuum, which seem to deny the possibility of fluid elements, i.e., hunks of fluid that remain together and have "integrity" in the sense of continuing to contain the same matter during their evolutions. This and other problems with the continuum concept, in view of the "atoms" and the vacuum, were discussed in Section 8.1. The Liouville equation offers us an opportunity to check the results of those continuum derivations against the more solid ground of Newtonian atomistics. It should be possible to derive those fundamental equations of change from the Liouville equation since both represent fundamental laws and nothing else. Neither have anything to do with equilibrium, reproducibility and determinism at the fluid level, or irreversibility. They are aimed at the same side of fluid phenomena and they should give the same rules as far as they go. Hence a derivation of the fundamental fluid equations of change from the totally innocuous Liouville equation would be a substantial support for our sorting out the various kinds of equations, and for our view of the hypothetical fluid continuum as being consistent with the view of atoms in the vacuum.

If we can derive the fundamental equations of change for a fluid in real physical space from the Liouville equation of change in the particle state space, then we not only have the satisfaction described above but also have a valuable tool. In the consideration of <u>mixtures</u> of different chemical species, the fluid concept becomes confused to the point where fundamental laws are no longer sufficient guides. At the same time there is no problem with atoms of different types flying through the vacuum in all directions, past others of different types, or even interacting with others of different types. Thus the atomistic view can offer a substantial aid to the more confusing continuum view in the formulation of fundamental equations of change for mixtures (Hirschfelder, Curtiss, Bird, 1954, Chapter 7).

The fundamental equations of change can be derived from the Liouville equation in the one-particle approximation most easily by first noticing the "moments" M(x,t) of the velocity distribution defined by

$$M_n(x,t) = \int dv\, v^n f(x,v,t) \tag{9.107}$$

According to the Liouville equation (9.83), the equations of change for the moments are

$$\frac{\partial}{\partial t} M_n = \int dv\, v^n \frac{\partial f}{\partial t} = - \int dv\, v^{n+1} \frac{\partial f}{\partial x}$$

$$\frac{\partial}{\partial t} M_n = - \frac{\partial}{\partial x} M_{n+1} \tag{9.108}$$

We can derive the fundamental fluid equations of change from these. The only difficulty in the derivation is due to the distinction between the fluid velocity v_0 (average molecular velocity) referred to in fluid dynamics as just v, as in (8.2) - (8.5), and the velocity v of a particular molecule relative to v_0. Some fluid properties of interest are determined by the difference

$$V = v - v_0 \tag{9.109}$$

which is often called the "peculiar velocity" of a molecule. We will call it the <u>relative</u> velocity. It is the velocity relative to an observer moving with the average fluid velocity $v_0(x,t)$. Thus the molecular velocities V have nothing to do with "convection" (moving with the fluid) but are strictly "conductive" (relative to the convective average v_0). We need to see any molecular velocity as a collective average v_0 plus a random (invididual velocity) V. Only then can we understand the "thermal" behavior of fluids.

Our first step in the translation of the Liouville moment equations (9.108) into the fluid equations of change (8.2) - (8.5) is to provide the relation between molecular properties and fluid properties. Suppose ϕ is any molecular such as one of the following

Mass	$\phi = m$
Velocity	$\phi = v$
Relative velocity	$\phi = V$
Relative momentum	$\phi = mV$
Relative kinetic energy	$\phi = \frac{1}{2} mV^2$ (9.110)

Notice that ϕ is a function of v or V. The density of ϕ at the fluid level is the fluid property P(x,t) defined by

$$P_\phi(x,t) = \int dv\ \phi(v) f(x,v,t) \tag{9.111}$$

We also have the number density

$$n(x,t) = \int dv\ f(x,v,t) \tag{9.112}$$

The total flux of any property ϕ is

$$J_\phi(x,t) = \int dv\, v\phi f \tag{9.113}$$

We can use the fluid velocity v_0 to separate the total flux into a convective contribution (carried by the moving fluid) and a conductive contribution which would be seen by an observer moving with the fluid,

$$v_o(x,t) = \frac{1}{n} \int dv\, v f \tag{9.114}$$

$$J_\phi = \int dv\, [v_o + (v - v_o)]\, \phi f = v_o P_\phi + J^*_\phi \tag{9.115}$$

The first term on the right side of (9.115) is the convective flux and the second term is the conductive flux:

$$J^*_\phi = \int dv (v - v_o) \phi f = \int dV\, V\phi f \tag{9.116}$$

In particular, we have the fluid velocity, denoted by v_0 in (8.2) - (8.5), as (9.114), and we have the mass density as

$$\rho = mn = m \int dv\, f \tag{9.117}$$

The momentum density is ρv_0 with convective flux $\rho v_0 v_0$. The pressure is the conductive flux of relative momentum

$$p = m \int dv\, VV f \tag{9.118}$$

for pressure is measured as a force per unit area (rate of momentum flow) by an observer moving with the fluid. The density of internal energy u(x,t) in the present context is the density of relative molecular kinetic energy

$$u = \int dv\, \frac{1}{2} mV^2 f \tag{9.119}$$

and the conductive heat flux is the relative flux of the relative kinetic energy,

$$q = \int dv\, \frac{1}{2} mV^2 V f \tag{9.120}$$

Let us terminate the listing here, for the moment, and check out some of the fundamental fluid equations of change. First we formulate the fluid properties in terms of the moments (9.107) and then we notice whether or not the moment equations of change (9.108) agree with the fluid equations of change. The fluid property ρ is clearly mM_0 where M_0 is the zeroth moment of f, and we have, from (9.108),

$$\frac{\partial}{\partial t} \rho = m \frac{\partial}{\partial t} M_o = - m \frac{\partial}{\partial x} M_1$$

$$\frac{\partial}{\partial t} \rho = - m \frac{\partial}{\partial x} (M_o v_o) = - \frac{\partial}{\partial x} (\rho v_o) \tag{9.121}$$

We see that the fluid equation of continuity (8.2) is implied by the Liouville equation.

Next we look for the fluid equation (8.3). The momentum density is

$$\rho v_o = mM_o(M_1/M_o) = mM_1 \tag{9.122}$$

Using its time derivative, according to (9.108), we have

$$\frac{\partial}{\partial t}\rho v_o = m\frac{\partial}{\partial t}M_1 = -m\frac{\partial}{\partial x}M_2 \tag{9.123}$$

Notice that the second moment M_2 is given by

$$\begin{aligned} M_2 &= \int dv\, v^2 f = \int dV (v_o + V)^2 f \\ &= \sigma_o^2 \int dV\, f + \int dV\, V^2 f \end{aligned} \tag{9.124}$$

A term disappeared because the sum of deviations from the average must be zero,

$$\int dV\, V f = 0 \tag{9.125}$$

As for replacing dv by dV in these integrals, that follows from (9.109) since the fluid velocity $v_o(x,t)$ is not a function of v. So we have (9.123) with (9.118) as

$$\frac{\partial}{\partial t}(\rho v_o) = -\frac{\partial}{\partial x}(\rho v_o^2) - \frac{\partial}{\partial x}P \tag{9.126}$$

This is the same as (8.3) so we may conclude that (8.3) is consistent with the Liouville equation.

By similar steps we can confirm the third fundamental fluid equation (8.4) as an expression of the Liouville equation. The internal energy density and the heat flux (conductive) are given by (9.119) and (9.120) and can be expressed in terms of the moments (9.107). Using the moment equations of change (9.108) we can derive the energy equation (8.4). The details are left to the reader. The conclusion is that the fundamental fluid equations of change are derivable directly from Newton's laws of motion, through the Liouville equation, and are, as argued above, strictly Newtonian. So far, in this application of the Liouville equations to fluid equations of change, we have no surprises.

But when we come to the entropy equation of change we encounter a large surprise. We start out to derive (8.5) as the equation of change for entropy, knowing that this equation contains none of the mysteries about entropy but merely expresses the second law of thermodynamics. There is nothing deterministic or irreversible about this equation, and it should come directly from the Liouville equation. But there is a problem right at the beginning of our derivation. What is the molecular property ϕ for which the entropy density can be given by (9.111), and the fluxes given by (9.113) and (9.116)? If we do not know that property, there is no way to relate (8.5) to the Newtonian-Liouville equation. Suddenly we are puzzled. Maybe the second law of thermodynamics does, after all, deal with mysteries beyond Newtonian mechanics.

Actually there is a candidate for the molecular property ϕ corresponding to the macroscopic entropy density, and that is $\ln f$, or more precisely, $-k\ln f$,

$$s(x,t) = -k \int dv\, f \ln f \qquad (9.127)$$

This is the candidate that comes from Boltzmann's H theorem and is still regarded as the prime candidate for definitions of fluid entropy. But this candidate produces serious problems, which are reviewed below.

First of all, $\ln f$ is not a macroscopic fluid property and there is no way to use that in the scheme outlined above to find a fluid equation of change. This could, of course, be part of the mystery about entropy that we are trying to dispel, so we do have a challenge here, even though we are only trying to justify (8.5) without yet getting into the heart of irreversibility as we see it.

Secondly, the definition (9.127) leads to a strange conclusion that has driven philosophers away from entropy as a signal of time anisotropy (Zenzen, 1977). We want to call them back, saying that (9.127) is wrong. Even Boltzmann knew it was wrong, but never became interested in that particular controversy. (There were too many other controversies.)

If the molecular property ϕ for s is $-k\ln f$, then (9.111) is (9.127), and we have the formula (9.127) for the density of entropy, and we can find the formula for the flux of entropy from (9.113) expressed as

$$J_s = -k \int dv\, v f \ln f \qquad (9.128)$$

where J_s is the flux of entropy as measured by a fixed observer. We have no formula for the conductive flux but can use (9.128) to calculate rate of creation or destruction of entropy as

$$\frac{\partial s}{\partial t} + \frac{\partial}{\partial x} J_s = \sigma \qquad (9.129)$$

$$\sigma = -k \int dv \ln f \left(\frac{\partial f}{\partial t} + v \frac{\partial f}{\partial x}\right) \qquad (9.130)$$

According to the Liouville equation (9.83), the right side of (9.130) is zero and we must conclude that entropy cannot be created or destroyed in our collisionless gas. Furthermore, we cannot extract the entropy equation (8.5) from the Liouville equation.

The impossibility of entropy creation as implied by (9.130) has led to errors in discussions about entropy. For example, the constancy of entropy has been rationalized on the ground that the sorting of velocities leads to a decrease in disorder which compensates the increase due to spatial expansion. That argument is discussed in more detail by Zenzen (1977). There is also support in the constancy of entropy for arguments that entropy creation and dissipation are caused by molecular collisions and cannot occur in a collisionless gas.

To clear up the problem that arises here we must find the correct definition of entropy. For that we start with the general Gibbsian definition of entropy

$$S = k \ln \Omega \qquad (9.131)$$

where Ω is the measure of the set of accessible (consistent with macroscopic data) states in the molecular state space, and that, in turn, can be estimated as W when and only when the distribution $\underline{n}$ is most probable relative to the macroscopic data, and only the current data, at that, (not past data). So the function f in (9.127) should not be the evolving f, as it should be in all other formulas like (9.111) and (9.113), but rather the most probable f relative to current fluid properties. That most-probable-in-a-current-sense f may have very little similarity to the actual evolving f, but nevertheless it is the f to be used in the definition (9.127) for the entropy density.

If we are ready to accept that definition of entropy and use the most probable f as given by (9.73) even though n and T may vary with x and t, then we can put (9.73) into (9.127) and find everything we need to derive (8.5) as the equation of change for entropy. If we have a moving fluid, then we have to modify (9.73) to

$$f = n\left(\frac{m}{2\pi kT}\right)^{1/2} e^{-\frac{m}{2kT}(v-v_o)^2} \qquad (9.132)$$

We see that we can derive the fundamental fluid equations of change from the Liouville equation, even the entropy equation if we define entropy properly. This confirms our claim that those fluid equations are strictly Newtonian and it gives us a new insight into how to define entropy. Generalizing to three dimensions would not change our conclusion here.

Now let us approach the more formidable equations of flux. There is no way they can come directly from Newtonian mechanics. There is nothing in Newtonian mechanics which could suggest the linear-in-gradient forms of (8.65), (8.66), (8.70), (8.71) summarized by (8.73) and generalized by (8.79). We can spin our wheels looking for such an approach, but let us acknowledge the only way ever found to make the approach. That is to assume that the system is in a most probable condition determined by a recent equilibrium. Let us assume, then, that our system does have the distribution (9.132) at the initial time when we "open the gates" and let something happen. According to the Liouville equation (9.83) the evolving f has a formula prescribed by (9.92), which is, for short times,

$$f(x,v,t) = f(x,v,0) - tv\frac{\partial}{\partial x} f(x,v,0) \qquad (9.133)$$

with

$$\frac{\partial}{\partial x} f(x,v,0) = \frac{\partial f}{\partial n}\frac{\partial n}{\partial x} + \frac{\partial f}{\partial T}\frac{\partial T}{\partial x} + \frac{\partial f}{\partial v_o}\frac{\partial v_o}{\partial x} \qquad (9.134)$$

where the f derivative in each term on the right side is a derivative of (9.132),

$$\frac{\partial f}{\partial n} = \frac{f}{n} \tag{9.135}$$

$$\frac{\partial f}{\partial T} = \left[-\frac{1}{2T} + \frac{m}{2kT^2} \right] f \tag{9.136}$$

$$\frac{\partial f}{\partial v_o} = \frac{m}{kT} (v - v_o) f \tag{9.137}$$

In view of (9.133) we have the conductive flux of any property ϕ as

$$J_\phi = \int dV\, V\phi \left(f_o - tv \frac{\partial}{\partial x} f_o \right) \tag{9.138}$$

where f_o is the "recent" (often called "local") equilibrium f (9.132) which can be expressed as

$$f_o = n \left(\frac{m}{2\pi kT} \right)^{1/2} e^{-\frac{m}{2kT} V^2} \tag{9.139}$$

It is clear that the first term on the right side of (9.138) is zero for any ϕ which is an even function of V, and this gives us the expected zero values for fluxes at equilibrium in all interesting cases except one. That one is the case where ϕ is the relative momentum and its flux at equilibrium is the pressure, which is not zero. Ignoring that exception for the moment, we keep only the second term on the right side of (9.138),

$$J_\phi = - \int dV\, V\phi\, tv \frac{\partial}{\partial x} f_o \tag{9.140}$$

and we can insert (9.134) to obtain, with the help of (9.135) - (9.137),

$$J_\phi = - L_{\phi 1} \frac{\partial n}{\partial x} - L_{\phi 2} \frac{\partial T}{\partial x} - L_{\phi 3} \frac{\partial v_o}{\partial x} \tag{9.141}$$

where

$$L_{\phi 1} = \int dV\, V\phi\, tv(f_o/n) \tag{9.142}$$

$$L_{\phi 2} = \int dV\, V\phi\, tv\left(-\frac{1}{2T} + \frac{mV^2}{2kT^2}\right) f_o \tag{9.143}$$

$$L_{\phi 3} = \int dV\, V\phi\, tv\, \frac{m}{kT}\, Vf_o \tag{9.144}$$

Notice that our use here of the Liouville equation in (9.138) has given us the fluxes in the form of a linear combination of gradients of fluid properties (9.141), which is very gratifying, even surprising, for that is the form of the successful stochastic equations of flux (8.78). In fact, we got the more general form of (8.79), which was more than we wanted at this stage. That form involving gradients of fluid properties came out of the Liouville equation only because we took f_o to be the most probable distribution (9.139) as determined by current values of fluid properties n, T, and v_o. We have incurred here a severe debt to the recent equilibrium.

And now let us force our way back to simple equations of flux, planning to return shortly to the complex equation (9.141). Suppose, for example, that the molecular property ϕ is the mass m, so the fluid property P is the density and the flux J is the flux of mass in Fick's equation (8.65), repeated here,

$$J_m = - D \frac{\partial}{\partial x} \rho \tag{9.145}$$

We can get this from (9.141) by including only the first term on the right side, ignoring, for the moment, contributions to diffusion due to gradients ("forces") other than the density gradient (the "primary" force for diffusion). Then we have the diffusion coefficient D as the first coefficient on the right side of (9.141), namely, (9.142):

$$D = m \int dV\, Vtv(f_o/n) \tag{9.146}$$

We will find in the arguments below that the absolute velocity v can be replaced by the relative velocity V without changing the value of D. (Such replacements do not always work, but this one does.) And so we have

$$D = m \int dV\, VVt(f_o/n) \qquad (9.147)$$

What we are referring to here as diffusion is, of course, "self-diffusion," for we have only one species of molecules diffusing through its own kind and not two different species diffusing through each other. The self-diffusion coefficient (9.147) can be evaluated and it is proportional to time (Biolsi, 1974).

If the gas is truly so sparse as to be collisionless (rarefied), then that is the end of our story for diffusion. But there is another case that deserves our attention. That is the case of the dilute-but-collision-dominated gases of "kinetic theory," where the molecules do collide, and very often, but over such short times that we can ignore the potential energy of interactions compared to the kinetic energy of motions at any one time. This is the exact predicament, for example, of a gas of rigid spheres (billiard balls).

In the case of the dilute-but-collision-dominated gas, which is the familiar case (atmospheric pressure), we can estimate transport properties by thinking of fluxes which grow (accelerate) from zero to terminal-velocity levels as we "open the gates" and let the gas leave equilibrium in a way that is very quickly brought to "terminal velocity" by the "friction" of collisions. We see the same thing when the ball accelerating downwards in a pin-ball machine is quickly brought to fairly constant speed down the board. Or when electrons moving through a metal are brought almost instantaneously to a fairly constant speed by collisions with ions. Or when the parachutist is brought quickly to a fairly constant speed by molecular collisions with atmospheric molecules.

With this terminal velocity concept in mind, we can estimate the time t in (9.145) as the average or typical time between collisions, the time when "terminal velocity" for flux acceleration is reached, and get an approximation for the diffusion coefficient (and other transport coefficients). To estimate the average time between collisions we first estimate the frequency of collisions as experienced by one molecule in a most probable (local equilibrium) collection of molecules. There is no other way to make that estimation. (Notice the subtle way that equilibrium, recent or local, manifests itself in these stochastic derivations.)

To estimate the frequency of collisions of one molecule we picture that one sphere of diameter σ moving with speed v during time t. It will collide with every other molecule in a cylinder of length vt and cross-sectional area $\pi\sigma^2$. The volume of that cylinder is $vt\pi\sigma^2$ and the number of the other molecules in it is the volume multiplied by the number density n. So the number of collisions of the molecule we follow with other molecules is $vt\pi\sigma^2 n$ and the number of collisions per unit time (divide by t) is

$$z = v\pi\sigma^2 n \tag{9.148}$$

Thus the time between collisions τ is the reciprocal of the frequency z,

$$\tau = \frac{1}{\pi\sigma^2 nv} \tag{9.149}$$

Incidentally, the distance traveled by the molecule between collisions is estimated here as

$$\lambda = \tau v = \frac{1}{\pi\sigma^2 n} \tag{9.150}$$

We can now put (9.149) into (9.146) and estimate the diffusion coefficient for the case of the typical gas ($t = \tau$).

There are some vague notions in the derivation of (9.149). Was that thinking in one or three dimensions? What is the speed v, the average speed? Actually, the derivation could be "one dimensional" if we visualize all the molecules in three dimensions moving in one direction. Or it could be three dimensional. The crux of the matter is the second question: what is v? Some of these questions are "nuisances" against the attempt here to express the main idea about how much the theory of transport phenomena depends on the presumption that the system is leaving equilibrium. But the question about v is interesting in itself, and does ultimately contribute a clue for us. So we will pursue some of the details.

First of all, let us move to three dimensions. The three-dimensional version of f_0 in (9.139) is

$$f_o = n\left(\frac{m}{2\pi kT}\right)^{3/2} e^{-\frac{m}{2kT}V^2}$$

$$\underline{V} = \underline{v} - \underline{v}_o(\underline{r}, t) \tag{9.151}$$

In the place of (9.140) we have

$$\underline{J}_\phi = - \int d\underline{V}\, \underline{V}\phi\, t\underline{v} \cdot \frac{\partial}{\partial \underline{r}} f_o \tag{9.152}$$

and for (9.145) and (9.147) we have

$$\underline{J}_m = - \underline{\underline{D}} \cdot \frac{\partial}{\partial \underline{r}} \rho \tag{9.153}$$

$$\underline{\underline{D}} = m \int d\underline{V}\; \underline{V}\,\underline{V}\, t(f_o/n) \tag{9.154}$$

where the vector notation(underlinings and dots) is explained by (8.35) and (8.46). The diffusion tensor (9.154) is a 3×3 matrix which is diagonal with

$$D = D_{xx} = D_{yy} = D_{zz}$$

We will use spherical coordinates (V, θ, φ) to evaluate D_{zz} with $V_z = V \cos\theta$ and

$$\int d\underline{V} = \int_0^{2\pi} d\varphi \int_0^{\pi} d\theta \sin\theta \int_0^{\infty} dV\, V^2$$

Thus we have

$$D = m \int_0^{2\pi} d\varphi \int_0^{\pi} d\theta \sin\theta \int_0^{\infty} dV\, V^4 \cos^2\theta\; t(f_o/n) \tag{9.155}$$

The interesting part is the scalar diffusion coefficient D in (9.155) which is essentially the same as the scalar defined in (9.147). The change to spherical coordinates

$(x,y,z) \to (r,\theta,\phi)$ for the evaluation of (9.155) was a matter of convenience in preparation for the various possible formulas for t, as we will see.

Secondly, we face the question of what is v in (9.149)? The first thought is that v should be the average speed of a molecule at the local equilibrium described by (9.151):

$$\bar{v} = \frac{1}{n} \int d\underline{V}\, V f_o$$

$$= \frac{1}{n} \int_0^{2\pi} d\phi \int_0^{\pi} d\theta \sin\theta \int_0^{\infty} dV\, V^3 n \left(\frac{m}{2\pi kT}\right)^{3/2} e^{-\frac{m}{2kT} V^2}$$

$$= (2\pi)(2)\left(\frac{m}{2\pi kT}\right)^{3/2} = \sqrt{\frac{8kT}{\pi m}} \qquad (9.156)$$

This is the famous formula for the average molecular speed at equilibrium, or the average relative molecular speed in a moving fluid at local equilibrium. But then we realize that the average speed we need for (9.148), if we want to use an average at all, is the average relative speed. After all, those spheres in the cylinder were not just sitting there waiting for the sweep of the sphere we followed but were moving in their own right. The average relative speed can be calculated for the local equilibrium distribution of velocities without much difficulty and it turns out to be the average speed formulated as in (9.156) with the mass replaced by the reduced mass, and that leads to (9.156) multiplied by $\sqrt{2}$. Going this route we end up with the basic three equilibrium properties, listed in all texts on Physical Chemistry and Kinetic Theory of Gases,

$$z = \sqrt{2}\pi\sigma^2 n\bar{v} \qquad \tau = \frac{1}{z}$$

$$\lambda = \bar{v}\tau = \frac{1}{\sqrt{2}\pi\sigma^2 n} \qquad (9.157)$$

These are referred to as the "mean" (meaning average) collision frequency, the mean collision time, and the mean free path.

If we use the mean collision time (9.157) for t in (9.155) we arrive at

$$D = a\lambda\bar{v} \tag{9.158}$$

where a is $\pi/8$. This offers an improvement over the traditional "mean-free path" theories which give $a = 1/2$. The attempt to offer the improvement without using the Liouville equation explicitly, as in (9.138) leads to $a = 1/3$ (Atkins, 1980). If we simply use $\bar{V}$ (magnitude) for v in (9.149) we find $a = \sqrt{2}/6$, and that is getting closer to the value that comes from the more rigorous theory involving the two-particle approximation (next section), namely, $a = 3/16$. We can come even closer if we let the speed v in (9.149) be the relative free-path speed found by considering the velocity of the molecule we follow relative to the velocities of the molecules it hits and averaging over those target molecules (Petersen, 1984).

The treatment of diffusion as described above can be applied to other transport phenomena, and we will give an outline of that below. But here it must be pointed out that the one-particle approximation described above has certain advantages over the traditional and familiar "mean-free-path" theories as described in any text on Physical Chemistry or Kinetic Theory. The method described above is not better in the sense of coming closer to a two-body approximation. It derives its advantages from the explicit connection to the Principle of Recent Equilibrium. We admit at the start that the only way to predict rates of irreversible processes in any approximation is to recognize the recent equilibrium and watch the system leave that equilibrium. The leaving can be described by the Liouville equation which provides the integral formula (9.140) for the fluxes. Since the formula involves an integration over molecular velocities it is not necessary to depend on mean free paths, mean speeds, those overly averaged properties used in the intuitive approaches to the one-particle approximation. Furthermore, we can, if we wish, abandon the limitation to rigid spheres and replace the collision cross section $\pi\sigma^2$ in the integral by cross sections that depend on velocity (or energy, for quantum mechanical models). We can come close to competing with the two-particle approximation. The most important advantage of the method described above is evident in (9.141), where a single flux is presented as a sum of contributions including contributions from "alien" forces. For example, we could repeat the derivation described above for the diffusion flux where only the first term on the right side of (9.141) was used, and replace that first term by the second term. Then we would come out with diffusion flux due not to the density gradient (Fick's law), but the temperature gradient. This is the "thermal diffusion" (the Soret effect) observed in liquids by Soret in 1880 but not expected in gases until predicted by Chapman (1917) using the two-particle approximation. And if we kept only the third term on the right side of (9.141) we would find a zero contribution to diffusion from a velocity gradient; there is no coupling between diffusion and viscosity (Hirschfelder, 1954, Section 11.2).

Returning to the simple equations of flux and using the recent equilibrium distribution f_0 with help from the Liouville equation for one particle, we now focus on the transport of heat for simple particles that have kinetic energy. The molecular property in (9.140) is

$$\phi = \frac{1}{2}\, mV^2 \qquad (9.159)$$

The conductive flux is given by (9.140) and (9.141) and we choose to keep only the middle term on the right side of (9.141). That is the term involving a temperature gradient, so it will lead us to the Fourier law (8.66) and give us the thermal conductivity $\varkappa$ as (9.143).

If we generalize to three dimensions after the manner used above for diffusion described by Eqs.(9.151) to (9.154), we find the thermal conductivity tensor

$$\underline{\underline{\varkappa}} = \int d\underline{V} \left(\frac{1}{2}\, mV^2\right) t\underline{V}\,\underline{V}\left(\frac{mV^2}{2kT^2} - \frac{3}{2T}\right) f_o \qquad (9.160)$$

which is diagonal. The diagonal elements are equal and they are

$$\varkappa = \frac{m}{2}\left(\frac{m}{2\pi kT}\right)^{3/2} \int d\underline{V}\, tV^4 \cos^2\theta\left(\frac{mV^2}{2kT^2} - \frac{3}{2T}\right) e^{-\frac{m}{2kT}V^2} \qquad (9.161)$$

The integration here is similar to the integration involved in (9.155).

The only further input needed for the evaluation of (9.161) is the selection of the time t. If we take t to be that actual time t, then we have the coefficient of thermal conductivity for a rarefied gas. If we seek the estimation for dilute-but-collision-dominated gases, with "terminal-velocity" fluxes, we could take t to be the mean collision time (9.157) or the free collision time (not averaged over velocities) or the "relative free collision time," described above. In any case, the result turns out to have the form

$$\varkappa = bkn\lambda\bar{v} \tag{9.162}$$

where k is the Boltzmann constant, λ is the mean free path (9.157), and v is the average velocity (9.156). The numerical coefficient b is 3/4 in the traditional mean-free-path theory and is corrected by intuitive arguments to 1/2 in modern texts. The value of b from (9.161) with t as the not-averaged time between collisions is $5\pi/8$ or 1.964 and the more sophisticated value from the two-particle approximation is $75\pi/128$ or 1.841. Here again we have an improvement for the one-particle approximation when we base it directly on the recent equilibrium distribution f and watch the system leave equilibrium.

The third simple equation of flux is Newton's formula for the transport of momentum (8.72) where the transport coefficient is the coefficient of viscosity η. Using our assumption of recent equilibrium (9.151) and using the Liouville equation to measure the growth of fluxes in a withdrawal from that equilibrium (9.138), we are led once again to (9.140) where this time the molecular property ϕ is the relative momentum

$$\phi = mv \; . \tag{9.163}$$

We single out the last term on the right side of (9.141) and we have the viscosity coefficients as (9.144) with (9.163).

If we move to three dimensions we have some complications in the notation. The flux of momentum is not a vector but rather is the pressure tensor of (8.31) with the off-diagonal elements indicated by (8.72). The flux equation comes out to be

$$\underline{\underline{p}} = -\underline{\underline{\underline{L}}} : \frac{\partial}{\partial \underline{r}} \underline{v}_o \tag{9.164}$$

with (9.144) generalized to a fourth-order tensor,

$$\underline{\underline{\underline{L}}} = \frac{m^2}{kT} \int d\underline{V}\ \underline{V}\,\underline{V}\,\underline{V}\,\underline{V}\, t f_o \tag{9.165}$$

There are 81 elements L_{ijkl} in this tensor, but we can quickly reduce the number of integrals to be evaluated by noticing that t (any choice) and f_0 depend only on the magnitude of V. The integral will be zero unless the array $V_i V_j V_k V_\ell$ is an "even" function of each of the components V_x, V_y, and V_z. So the four indices i, j, k, ℓ must all be the same, or else they must consist of two pairs of equal indices. All integrals of each type are undoubtedly the same, so we have only two integrals to evaluate.

Rather than pursue that notational problem to the end let us get quicker to the point by stipulating that the fluid is moving in the y-direction with a gradient of v_{oy} in the z-direction, as pictured in Fig.8.6. Then the only nonzero component of the gradient tensor is

$$\left(\frac{\partial}{\partial \underline{r}} \underline{v}_o\right)_{32} = \frac{\partial}{\partial x_3} v_{02} = \frac{\partial}{\partial z} v_{oy} \tag{9.166}$$

Hence the ij component of (9.164) is

$$p_{ij} = L_{ij23} \frac{\partial}{\partial x_3} v_{02} \tag{9.167}$$

where i,j must be 2,3 or 3,2. Actually they are the same. So we have (8.72) with the viscosity coefficient η as L_{2323},

$$\eta = \frac{m^2}{kT} \int d\underline{V}\ V_y^2 V_z^2\, t f_o \tag{9.168}$$

Using spherical coordinates for the integration, as above, we can find η for the various choices of t. Keeping t as just the time t, independent of the molecular velocities, we find the viscosity coefficient (proportional to time) for the rarefied gas. For the collision-dominated gas, we find the form

$$\eta = cmn\lambda\bar{v} \tag{9.169}$$

with a numerical coefficient c that depends on the choice of t. The crudest mean-free-path theory gives $c = 1/2$, which is the best value from any of these one-particle approximations considered here, for the more rigorous value from the two-particle approximation is .499. The traditional approach as improved in modern texts gives $c = 1/3$. If we choose t in a way that makes use of the advantages of the present formulation, we find $c = (2/3)\sqrt{2/\pi} = .532$. Experimental values at common temperatures for simple molecules go from .361 for CO_2 to .665 for He. To argue for or against any of the methods referred to here is certainly a matter of nitpicking.

Actually the criticism of nitpicking could also be directed against the earlier results we obtained for diffusion and thermal conductivity. If all we care about is the predicted numerical value, then none of these one-particle approximations is as good as the two-particle approximation. This is especially evident in the case of viscosity, and that is most embarassing because the molecular diameters used in most estimations of the type we consider here come from viscosity data (the easiest to find experimentally). But let us not miss the forest for the trees. The gradient forms of the simple equations of flux are implied in a logical manner by the use of the Liouville equation (9.138) (Newtonian) as applied to recent equilibrium (9.139). The arguments used to get that form in mean-free-path theories seemed quite arbitrary and suspect, even though they did provide the same remarkable result. We will find a similar reduction to logical and systematic procedures without any non-Newtonian mysteries when we acknowledge the role of recent equilibrium in the two-particle approximation described in Section 9.7.

9.6 Complex Equations of Flux in the One-Particle Approximation

In the previous section we observed that equations of equilibrium have little debt to mechanics compared to their debt to statistics, and that fundamental equations of change have no debt to statistics. In the case of the simple flux equations, we found that they have a huge debt to statistics in the sense that they hold only in the wake of randomness (equilibrium) where statistics prevails. The deliberate use of mechanics to describe the evolution of the system during early stages in the wake of randomness offered advantages over the stochastic arguments of past derivations, and now we focus on the most important of those advantages. We refer here to the description of the coupling of transport phenomena known as Irreversible Thermodynamics, or the thermodynamics of irreversible processes.

We could start by going back to (9.141) and exploring the coupling implied by that equation between the various "forces" (gradients of fluid properties) and "fluxes" (transport rates for the conduction of various properties). But such a discussion would be based on the choices of constraints leading to (9.139) as the most probable distribution of molecular velocities. This distribution is not necessarily the most probable distribution. What distribution is most probable depends on the "constraints." So let us be very general here and not say which constraints we have chosen. They will be identified later.

Going back to the drawing board, so to speak, we go back to W in (9.8) and ask for the most probable distribution $\underline{n}$ subject to constraints like (9.13) and (9.28) or (9.40). In other words, we insist on (9.18) but leave open what constraining conditions must hold. In fact, we replace those earlier constraints by new ones which are abstract.

Let us say, for example, that we have macroscopic properties A, B, C, ... which have known (fixed) values, and they are sums of contributions from individual molecules. This is the case, for example, for mass, momentum, angular momentum, kinetic energy, and other properties. Thus we have, by analogy with (9.40) and (9.42),

$$A = \sum_s n_s a_s$$

$$B = \sum_s n_s b_s$$

$$C = \sum_s n_s c_s$$

$$\cdots \qquad (9.170)$$

and we consider only variations dn_s which satisfy

$$dA = \sum_s a_s \, dn_s = 0$$

$$dB = \sum_s b_s \, dn_s = 0$$

$$dC = \sum_s c_s \, dn_s = 0$$

$$\cdots\cdot \qquad (9.171)$$

When we look for the n_s that satisfy (9.18) for all dn_s that satisfy (9.171) and use the method of Lagrange multipliers in the same way we did above in Section 9.1, we find the most probable distribution n_s given by a formula that depends on parameters (Lagrange multipliers), call them α, β, γ, ... for each of the constraints (9.170). Inevitably, by analogy with (9.31) or (9.44), we find the exponential form

$$n_s = ne^{-\alpha a_s - \beta b_s - \gamma c_s - \cdots} \qquad (9.172)$$

involving a linear combination of Lagrange multipliers, which is so typical and characteristic of the method of statistical mechanics. It provides very reliable predictions for the systems of interest here, with very large numbers of molecules, provided we can evaluate the multipliers (the statistical parameters). We will have to either eliminate them algebraically or welcome them as new independent variables whenever that can be justified.

Moving to the continuum of states, we find (9.172) expressed as

$$f = e^{-\alpha a_s - \beta b_s - \gamma c_s - \cdots} \qquad (9.173)$$

This switch is like the switch from (9.44) to (9.67) except that we do not try here to pull out a normalization constant A. We leave that for later determinations. And we do not express molecular properties in the relative velocity $\underline{V}$ but rather stick to $\underline{v}$ in order to avoid prejudice in the selections of properties. The transition to $\underline{V}$ for fluids where the fluid velocity $\underline{v}_0$ is a recognized property is easy to make. Descriptions of fluids that do not have that commitment are descriptions of stagnant fluids in which there can be many processes going on. They happen to be fluids that are not moving convectively.

Let us go back to (9.110) and say that $\phi_1(\underline{v})$ is any molecular property. Its fluid density is P_i, (9.111), repeated here,

$$P_i = \int d\underline{v}\,\phi_i f \tag{9.174}$$

and its flux is J_i, (9.113),

$$\underline{J}_i = \int d\underline{v}\,\underline{v}\phi_i f \tag{9.175}$$

The function f itself is determined by the P_i (9.174) to be the most probable distribution for the given P_i consistent with the form of (9.173),

$$f = e^{-\sum_i \alpha_i \phi_i} \tag{9.176}$$

We see the set of equations (9.174) as the set of equations to be solved for the α_i. The number of such equations is the same as the number of the unknown α_i. To be more explicit we express (9.176) as

$$f(\underline{r},\underline{v},t) = e^{-\sum_i \alpha_i(\underline{r},t)\phi_i(\underline{v})} \tag{9.177}$$

where the $\alpha_i(\underline{r},t)$ are the Lagrange multipliers determined at each point $\underline{r}$ and each time t from the values of the $P_i(\underline{r},t)$. Thus we find the most probable "current" (at the current time) distribution $f(\underline{r},\underline{v},t)$ over velocities $\underline{v}$. The separation of the dependence on $(\underline{r},t)$ at the fluid level, which is the argument of the functions α_i, and $\underline{v}$ at the molecular level, which is the argument of the ϕ_i, is very important in the discussion that follows.

The method just described for determining f is based entirely on statistical mechanics without any consideration of the evolution of f as prescribed by the Liouville equation. What we get by this method is the currently most probable f relative to the current values of the densities P_i without any regard for the question of whether this evolving f satisfies the Liouville equation. (It never does.) Let us label this currently most probable f by using an asterisk,

$$f^*(\underline{r},\underline{v},t) = e^{-\sum_i \alpha_i(\underline{r},t)\phi_i(\underline{v})} \tag{9.178}$$

where the $\alpha_i(\underline{r},t)$ are determined as described above from the $P_i(\underline{r},t)$. It may seem that we are over-emphasizing a trivial point, but our motivation will become evident quickly.

Let us now introduce another f, which is more important in most practical applications than the one described above. We refer to the evolving f determined by initial conditions and the Liouville equation. This is the f described above by (9.97)

$$f(\underline{r},\underline{v},t) = f(\underline{r}-\underline{v}t,\underline{v},0) \tag{9.97}$$

It was approximated for short times above by

$$f(\underline{r},\underline{v},t) = f_o - tv \cdot \frac{\partial}{\partial \underline{r}} f_o$$

$$f_o = f(\underline{r},\underline{v},0) \tag{9.179}$$

We assume that the initial value is determined by statistical mechanics

$$f_o = e^{-\sum_i \alpha_i(\underline{r},0)\phi_i(\underline{v})} \tag{9.180}$$

based on the initial values of the α_i as determined by the initial values of the P_i. Then we forget the P_i and let f evolve in a way that is determined at the microscopic (molecular) level by mechanics, i.e., by the Liouville equation, for a time t. At that point we can use the evolved f to predict evolved values of the P_i and J_i based on (9.174) and (9.175). If those predictions agree with the currently observed values of P_i and J_i (they do), then we will say that the method was successful. But nowhere in this method was there a role to be played by f^* (except at the initial time).

The role to be played by f^* does not appear in the predictions of evolved values for the P_i and J_i but it does make a dramatic appearance when we come to the predicted value for the entropy density $s(\underline{r},t)$. Unlike the other densities, which are predicted from f, the entropy density is predicted from f^*. If we accidentally use f we will obtain a density and a flux of entropy which will (1) be wrong, (2) imply no creation or destruction of entropy, (3) offer no molecular interpretation for the entropy equation of change (8.5), as we noticed above, (4) block our route to the understanding of the Onsager reciprocal relations (4.3), and (5) block our route to the understanding of the Boltzmann H-theorem. Much of the confusion around the issues we consider in this book come from a failure to make the distinction between f and f^*. The importance of that distinction does not wait for a consideration of more sophisticated models; it crops up right here, in the one-particle approximation.

We noticed above the problems that come from the erroneous definition of the entropy density and flux based on f in (9.128) and (9.129). Those formulas should be replaced by

$$s = -k \int d\underline{v}\; f \ln f^* \tag{9.181}$$

$$\underline{J}_s = -k \int d\underline{v}\, \underline{v} f \ln f^* \tag{9.182}$$

In fact, it would be more faithful to our argument to replace the first f on the right sides of (9.181) and (9.182) by f^*, but there it does not matter which f we use. The reason for that equivalence goes back to the formula for the P_i,

$$P_i = \int d\underline{v}\; \phi(\underline{v}) f \tag{9.183}$$

Here it does not matter whether we use f or f^*, for f is what is used to predict the P_i and then those P_i are used through (9.183) to predict f^*. (It does matter which f we use to predict the fluxes; it has to be the evolved f and not the current most probable f.) So we can view (9.181) as a P_i where the molecular property ϕ is $\ell n f^*$. This makes even more sense when we notice that $\ell n f^*$ really is a linear combination of molecular properties. From (9.176), now recognized as f^*, we have that linear combination as

$$\ell n\, f^* = -\sum_i \alpha_i(\underline{r}_i, t)\, \phi_i(\underline{v}) \tag{9.184}$$

At this point let us back up and take an overview of the method of nonequilibrium statistical mechanics. It may be true that we are rushing a bit to try to extract this overview at the level of the one-particle approximation, but actually the possibility for that extraction is in front of us and it suffers in no way from the level of the approximation. What we have is a scheme in which property densities P_i are given at an initial time $t = 0$ and used to find the most probable f, namely f^*, at the initial time. In that sense we start at the macroscopic fluid level with the P_i and descend to the microscopic molecular level to find f. Then we ignore what happens at the macroscopic fluid level and follow the evolving f at the molecular level for a time t. Then we use the evolved f to go back up and predict the new values of the P_i and the J_i at time t. (The J_i were initially zero.) That is the whole story if you leave entropy out of it. If you want to predict entropy at time t, then you cannot use the evolved f but must use the predicted P_i at time t to find f^* and then use that to predict the entropy. The part of this prescription that does not involve entropy is summarized in Fig.9.1.

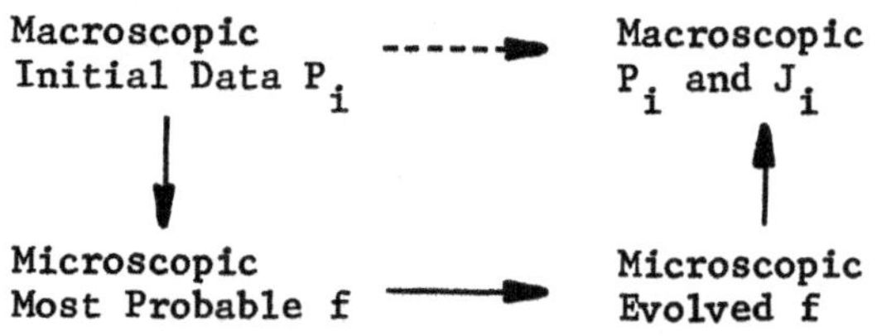

Fig.9.1 The Method of Nonequilibrium Statistical Mechanics

If we include the role of entropy in the summary presented by Fig.9.1, we must add some notes. The initial f is the same as f^* but the f at time t is not the same as f^* at time t. We can derive f^* at the later time from the evolved f, but to do that we must first evaluate the evolved P_i and then use those to find f^*. We see that f^* is really not a molecular property but a fluid property to be defined and evaluated at the fluid level. The same conclusion applies to entropy. This should be no surprise. All we are saying is that entropy is a macroscopic property with no microscopic counterpart.

Let us now look for the equations of flux implied by the general formulation of nonequilibrium statistical mechanics in the one-particle approximation. At the initial time (recent equilibrium) we have f as f_o in (9.180). At later times the evolved f is, according to (9.97),

$$f(\underline{r},\underline{v},t) = e^{-\sum_i \alpha_i(\underline{r}-\underline{v}t,0)\phi_i(\underline{v}_i)} \tag{9.185}$$

which is not the same as the most probable f of (9.178). We can also present the evolved f according to (9.179) as,

$$f = f_o - t\left(\underline{v} \cdot \frac{\partial}{\partial \underline{r}}\right) f_o = f_o + t\underline{v} \cdot \sum_i \frac{\partial \alpha_i}{\partial \underline{r}} \phi_i f_o \tag{9.186}$$

Ignoring the first term f_o, which disappears in calculations of flux, we have the flux J_j in (9.175) as

$$\underline{J}_j = \sum_i \left[\int d\underline{v}\, \underline{v}\,\underline{v}\, t\phi_i\phi_j f_o\right] \cdot \frac{\partial}{\partial \underline{r}} \alpha_i \tag{9.187}$$

Interchanging i and j, merely to satisfy prejudices for starting with earlier letters in the alphabet, we express (9.187) as

$$\underline{J}_i = \sum_i \underline{\underline{L}}_{ij} \cdot \underline{X}_j \tag{9.188}$$

where

$$\underline{X}_j = \frac{\partial}{\partial \underline{r}} \alpha_j \tag{9.189}$$

$$\underline{\underline{L}}_{ij} = \int d\underline{v}\ \underline{v}\,\underline{v}\ t\phi_j \phi_i f_o \tag{9.190}$$

The important feature of (9.187) to be noticed here is explicit in (9.190), namely the symmetry:

$$\underline{\underline{L}}_{ij} = \underline{\underline{L}}_{ji} \tag{9.191}$$

This symmetry goes even further. Looking at a particular tensor with given i and j, the tensor itself is symmetric because $\underline{v}\,\underline{v}$ has the symmetry $v_x v_y = v_y v_x$. In other words, if we abandon the collecting vector notation and considered each separate component of each vector and tensor, using integer labels, we can express (9.188) as (4.2) or (8.79)

$$J_i = \sum_j L_{ij} X_j \tag{9.192}$$

and conclude that the L matrix is symmetric,

$$L_{ij} = L_{ji} \tag{9.193}$$

If we treat the $\underline{X}_j$ (gradients of α_j) as "forces" (gradients of fluid properties which cause transport), then we have (9.191), and ultimately (9.193), as expressions of the Onsager reciprocal relations, i.e., the basic "law" of Non-equilibrium Thermodynamics (Prigogine, 1955).

We are tempted to conclude here that we have finally found the origin of the reciprocal relations, as requested in Section 8.7. We could say that those relations govern the way in which systems leave equilibrium and can be expected to hold only in situations where we deal with a fluid being forced to leave a recent equilibrium. That would add considerable support to our arguments about the importance of recent equilibrium in descriptions of non-equilibrium phenomena.

But there is one problem we face here. The theory of nonequilibrium thermodynamics coming from Onsager offers the symmetry (reciprocal) relations (9.193) only for the special case where the fluxes J_i and forces X_i have been defined in a certain way. In our situation, where the forces are abstract gradients of Lagrange multipliers, not yet identified in terms of fluid densities, it is not obvious that we have the right forces.

According to the recipe of nonequilibrium thermodynamics, the reciprocal relations (9.193) need hold only if the fluxes and forces are defined in such a way that the rate of creation of entropy density σ is given by

$$\sigma = \sum_i J_i X_i \tag{9.194}$$

This is the rate that was formulated above on the right side of (9.130), which turned out to be zero when we were using a wrong definition of entropy. Now that we have the right formula, we can check our results above against (9.194) to determine whether or not we do have, indeed, the Onsager reciprocal relations in (9.193).

To make this check we go back to (9.181) and (9.182), and recognize f^* as (9.178), omitting vector notations, and we find

$$s = k \sum_i P_i \alpha_i \tag{9.195}$$

$$J_s = k \sum_i J_i \alpha_i \tag{9.196}$$

Thus the absolute rate of entropy creation introduced in (9.130) is, in one dimension,

$$\sigma = \frac{\partial}{\partial t} s + \frac{\partial}{\partial x} J_s = \sum_i \alpha_i \left[\frac{\partial}{\partial t} P_i + \frac{\partial}{\partial x} J_i\right]$$

$$+ \sum_i \left[P_i \frac{\partial}{\partial t} \alpha_i + J_i \frac{\partial}{\partial x} \alpha_i\right] \tag{9.197}$$

The first term on the right side of (9.197) is zero according to (9.108). In fact, that is the only part of the entropy production (creation rate, density thereof) included above (9.130) where we used the wrong f in the evaluation of entropy and found no entropy production.

Now we have the entropy production as

$$\sigma = \sum_i P_i \frac{\partial}{\partial t} \alpha_i + \sum_i J_i \frac{\partial}{\partial x} \alpha_i \tag{9.198}$$

The first term on the right side is the rate of change of the number density n,

$$\frac{\partial}{\partial t} n = \frac{\partial}{\partial t} \int dvf = \frac{\partial}{\partial t} \int dr\, e^{-\sum_i \alpha_i \phi_i} = -\sum_i \frac{\partial \alpha_i}{\partial t} P_i \tag{9.199}$$

Since the spatial derivatives of the α_i are the "forces" X_i of (9.189), we have (9.198) as

$$\sigma = -\frac{\partial}{\partial t} n + \sum_i J_i X_i \tag{9.200}$$

If the fluid is stagnant, with no convective motion even though every other process is allowed, then we can invoke the equation of continuity (9.121) in the form,

$$\frac{\partial}{\partial t} n + \frac{\partial}{\partial x} (nv_o) = 0 \tag{9.201}$$

and conclude that the first term on the right side of (9.200) is zero. Furthermore, the fluxes J_i in the surviving term become the <u>conductive</u> fluxes, which are the only fluxes allowed in the thermodynamic formula (9.194). So we see that our selection of fluxes and forces does satisfy the thermodynamic criterion for the one-dimensional stagnant fluid, and for that case we can be sure that the relations (9.193) are the Onsager reciprocal relations.

It is easy to generalize the derivation above to three dimensions and to the case of a fluid with convective motion. We merely have to return to the vector notation of (9.175) and separate the total flux there into conductive and convective parts. It is also necessary to replace the $\phi_1(\underline{v})$ by $\phi_1(\underline{V})$ in (9.185) because the $\underline{V}$ depend through $\underline{v}_0$ on $\underline{r}$ and t. But the cancellations of terms that occurred in the derivation above are generalized in such a manner that the end result is, again, (9.194), where the $\underline{J}_1$ are purely conductive.

Our conclusion here is that the symmetry relations governing the couplings among transport processes are due to the recent equilibrium in the history of the fluid. There is no reason to expect these relations to hold in any predicament other than the case of a forced withdrawal from a young and recent equilibrium. Hence we may say that the cause of the symmetry is macroscopic reversibility, referring here to the inertial criterion for reversibility (criterion 1). There is no such reversibility at the microscopic level. It is the nearness to equilibrium (in past time) at the macroscopic level that justifies the derivation of the symmetry relations.

It is ironic that the traditional explanation of the symmetry relations finds their origin in microscopic reversibility, implying that there is no macroscopic reversibility, when, in fact, the origin is in macroscopic reversibility where there is no microscopic reversibility. More interesting, perhaps, than the origin of irreversibility, is the origin of that blunder. On the one hand, we know that such blunders are easy, and there are probably many in the text above, but on the other hand, this particular blunder has caused a great deal of confusion and inhibited the applications of statistical mechanics. It is, we may say, only a blunder in terminology, but it must be explored, for it has had considerable influence.

The idea involved in the traditional terminology is that irreversibility is a temporal thing (criterion 2) which dominates fluid behavior at the macroscopic level and never occurs in molecular behavior. That is wrong, as we have seen, for temporal irreversibility does not exist in any description at any level. To even suggest that kind of irreversibility is to deny the power of imagination to conceive reversed processes, which makes it impossible to understand the exclusion irreversibility of thermodynamics. With that idea in mind, however, it was possible to interpret the Onsager derivation of the symmetry relations as an unusual exposure of the reversible particle mechanics at the irreversible macroscopic level. Hence the reference to microscopic reversibility. No such reference would have been made if the macroscopic processes had not been labeled "irreversible." In retrospect we can say that the terminology was misleading. In fact, even if there were such a thing as temporal irreversibility at the macroscopic level the terminology would still be misleading, for the reversibility at the microscopic level would be just as prevalent in situations where the symmetry relations do not hold at the macroscopic level as in situations where they do hold. (Note 27)

There is a role for equilibrium in traditional treatments of nonequilibrium processes. For example, the symmetry relations and linear laws (9.192) are sometimes regarded as attributes of the asymptotic approach to a new and distant future equilibrium. And it is commonly assumed that the linearities and symmetries of transport theory and irreversible thermodynamics hold only near that equilibrium, where "local" equilibrium has "set in." Here <u>near</u> means that the spatial gradients (forces) are small, as if to say that we are sliding into the new equilibrium at sufficiently close range that we can ignore terms with higher derivatives and nonlinear terms. It is regarded as surprising that Fick's law or Fourier's law works so well at early times in diffusion and heat conductivity in situations where the gradients are nearly infinite. The fact is that all of the equations of flux are restricted to predicaments near equilibrium, but not <u>near</u> future uniformity in the sense of small spatial gradients. Rather <u>near</u> in the sense of nearness in <u>time</u> to a past equilibrium. If that equilibrium was young and recent, we have the assurance of randomization, and the forced withdrawal will be very orderly, reproducible, and predictable for a sufficient period of time. Otherwise, we have no base for predictions.

When we combine the false notion about an unusual emergence of microscopic reversibility at the macroscopic level with the misleading idea of nearness to equilibrium as a spatial thing, we obtain the current confusion about, and disinterest in, theories about irreversible processes. Certainly it is desirable to reconsider these notions and assumptions at the simplest and most fundamental level. There may be some useful applications.

9.7 The Two-Particle Approximation

Consider two particles of the same mass with a potential energy of interaction u(r), where r is the distance between the particles. The force on particle i due to particle j is

$$\underline{F}_i = -\frac{\partial u}{\partial \underline{r}_i} \tag{9.202}$$

Thus the acceleration is

$$\underline{a}_i = \frac{1}{m}\underline{F}_i = -\frac{1}{m}\frac{\partial u}{\partial \underline{r}_i} \tag{9.203}$$

Suppose we have a gas of particles paired in such a way that two interacting particles never interact with a third particle while interacting with each other. This could be a very dilute gas or a gas composed of molecules with very short-range forces. The best idealization is a gas of rigid spheres. In the limit that the rigid spheres become small, the model here becomes exact in the many-particle approximation.

The density of particles in the state space of $\underline{s} = (\underline{r}, \underline{v})$ is the distribution function $f(\underline{s}, t)$ of the previous sections. We will refer to f as the "one-particle distribution function" $f^{(1)}$. The density of pairs in the two-particle state space of $(\underline{s}_1, \underline{s}_2)$ is the two-particle distribution $f^{(2)}$. The function $f^{(2)}$ must be normalized in the following way:

$$f^{(1)}(\underline{s}_1, t) = \int d\underline{s}_2 f^{(2)}(\underline{s}_1, \underline{s}_2, t)$$

$$f^{(1)}(\underline{s}_2, t) = \int d\underline{s}_1 f^{(2)}(\underline{s}_1, \underline{s}_2, t) \tag{9.204}$$

If we think of $f^{(2)}$ as the density of a probability fluid on the two-particle state-space, then we have the equation of continuity as the Liouville equation (9.101),

$$\frac{\partial}{\partial t} f^{(2)} + L f^{(2)} = 0 \tag{9.205}$$

$$L = \underline{v}_1 \cdot \frac{\partial}{\partial \underline{r}_1} + \underline{v}_2 \cdot \frac{\partial}{\partial \underline{r}_2} + \underline{a}_1 \cdot \frac{\partial}{\partial \underline{v}_1} + \underline{a}_2 \cdot \frac{\partial}{\partial \underline{v}_2} \tag{9.206}$$

We also have the evolution (9.103) described by

$$f^{(2)}(\underline{s}_1, \underline{s}_2, t) = e^{-tL} f^{(2)}(\underline{s}_1, \underline{s}_2, 0) \tag{9.207}$$

In order to apply (9.207) to fluid dynamics we have to identify constraints, find a most probable distribution, and then watch the distribution evolve. A short-time evolution will be sufficient to provide the fluid rates of change if we can find a suitable initial condition.

The constraints are the fluid property values, the densities P_i and fluxes $\underline{J}_i$. Insofar as we can ignore properties ϕ_i that depend on both $\underline{s}_i$, such as potential energy, we can use the P_i and J_i to first find $f^{(1)}$ and then use $f^{(1)}$ to find $f^{(2)}$. (Here we have in mind rigid spheres, although the results we get can be generalized.) If we start at equilibrium we can ignore the fluxes $\underline{J}_i$. Then finding the most probable $f^{(1)}$ is exactly the same as in the one-particle approximation above. Our starting point here then is the following. Given $f^{(1)}$, find $f^{(2)}$ that is most probable among all candidates allowed by the constraints (9.204).

To solve this problem we return to the discrete distribution described by (9.8). If n_{ij} is the number of pairs with particle one in state i and particle two in state j, and N is the total number of pairs, then the number of ways to assign particles to states and get the distribution $\underline{n} = (n_{11}, n_{12}, \ldots n_{21}, \ldots)$ is

$$W = \frac{N!}{n_{11}!n_{12}! \ldots n_{21}! \ldots} \tag{9.208}$$

Thus the analog here of (9.18) is

$$-dF = \sum_{i,j} \left(\ln \frac{n_{ij}}{N}\right) dn_{ij} = 0 \tag{9.209}$$

The constraints (9.204) become

$$n_i = \sum_j n_{ij}$$
$$n_j = \sum_i n_{ij} \tag{9.210}$$

and hence the constraints on the variations dn_{ij} are

$$dn_i = \sum_j dn_{ij}$$

$$dn_j = \sum_i dn_{ij} \tag{9.211}$$

Following the method of Lagrange multipliers we multiply the first equation in (9.211) by α_i and the second by α_j, and add the results to (9.209) to obtain

$$\sum_{i,j} \left(\ell n \frac{n_{ij}}{N} + \alpha_i + \alpha_j \right) dn_{ij} = 0 \tag{9.212}$$

We choose the α_i to cause the coefficients to vanish and hence we find

$$n_{ij} = Ne^{-\alpha_i - \alpha_j} \tag{9.213}$$

To find α_i we use

$$n_i = \sum n_{ij} = Ne^{-\alpha_i} \sum_j e^{-\alpha_j} \tag{9.214}$$

Denoting the sum by Q we have

$$e^{-\alpha_i} = \frac{1}{NQ} n_i \tag{9.215}$$

This covers the cases of all values of i, so our most probable distribution turns out to be

$$n_{ij} = \frac{1}{NQ^2} n_i n_j$$

Normalization leads to the simple formula

$$n_{ij} = n_i n_j \tag{9.216}$$

Carrying this result over to our function $f^{(2)}$ we see that the most probable distribution is

$$f^{(2)*}(\underline{s}_1, \underline{s}_2, t) = f^{(1)}(\underline{s}_1, t) f^{(1)}(\underline{s}_2, t) \tag{9.217}$$

In other words, the two-particle distribution function <u>factors</u> into a product of one-particle distribution functions. That way of relating $f^{(2)}$ to $f^{(1)}$ is very simple and very important, but we have to remember that the factorization is not a general characteristic of $f^{(2)}$. It applies only to the <u>most probable</u> situation. Strictly speaking, we are limited to equilibrium.

If we had given more attention to the intermolecular force (longer range or larger spheres) then we would have obtained (9.217) in the corrected form

$$f^{(2)*} = Y f^{(1)} f^{(2)} \tag{9.218}$$

where Y is a function of $\underline{r}_1$ and $\underline{r}_2$ which goes to unity as the distance between the $\underline{r}_i$ becomes large. This is the function Enskog (1922) introduced. If we admit that the most probable situation prevails only at equilibrium, then we have (9.218) as the Gibbsian equilibrium function. It is well known that the Gibbs function does factor in the sense of (9.218).

Let us now consider the application of the mechanics of the Liouville equation and the statistics of the most probable $f^{(2)}$ to the prediction of fluid equations. For equations of equilibrium we can forget the mechanics and use only the most probable $f^{(2)}$. There is nothing new here to be found for equilibrium. The Gibbs Theory (Section 9.2) says it all.

For equations of change we can hopefully forget the statistics and use only the Liouville equation. Since the fluid properties we have in mind are "one-particle properties" in the sense of being derivable from $f^{(1)}$ as in (9.111) and (9.115), the way to use the two-particle Liouville equation here is to first derive a corrected form of the one-particle equation (9.83) and use that to find fluid equations of change.

To find the one-particle equation we start with the Liouville equation (9.205) and integrate it over the coordinates $\underline{s}_2 = (\underline{r}_2, \underline{v}_2)$. The first term becomes, according to (9.204)

$$\frac{\partial}{\partial t} \int d\underline{s}_2 f^{(2)}(\underline{s}_1, \underline{s}_2, t) = \frac{\partial}{\partial t} f^{(1)}(\underline{s}_1, t) \tag{9.219}$$

The other four terms come from (9.206):

$$\int d\underline{s}_s \underline{v}_1 \cdot \frac{\partial}{\partial \underline{r}_1} f^{(2)} = \underline{v}_1 \cdot \frac{\partial}{\partial \underline{r}_1} f^{(1)} \tag{9.220}$$

$$\int d\underline{s}_2 \underline{v}_s \cdot \frac{\partial}{\partial \underline{r}_2} f^{(2)} = 0 \tag{9.221}$$

$$\int d\underline{s}_2 \underline{a}_1 \cdot \frac{\partial}{\partial \underline{v}_1} f^{(2)} = - J \tag{9.222}$$

$$\int d\underline{s}_2 \underline{a}_2 \cdot \frac{\partial}{\partial \underline{v}_2} f^{(2)} = 0 \tag{9.223}$$

The zero values of (9.221) and (9.223) come from the boundary condition. In this context it is generally assumed that $f^{(2)}$ goes to zero at large $\underline{r}_1$ and $\underline{v}_1$, which implies

$$\int d\underline{r}_2 \frac{\partial}{\partial \underline{r}_2} f^{(2)} = \int d\underline{v}_2 \frac{\partial}{\partial \underline{v}_2} f^{(2)} = \underline{0} \tag{9.224}$$

The integral (9.222) is identified by J so that we can express the whole equation in the form

$$\frac{\partial}{\partial t} f^{(1)} + \underline{v}_1 \cdot \frac{\partial}{\partial \underline{r}_1} f^{(1)} = J \qquad (9.225)$$

The new feature in (9.225) not included in the one-particle Liouville equation (9.83) is the integral J on the right side. If we multiply (9.225) through by each molecular property $\phi_i(\underline{V}_1)$ and integrate over $\underline{v}_1$, we will obtain, after some manipulations, the fluid equations of change. And if we ignore J (i.e., set J = 0) we will obtain the same equations we found in the one-particle approximation. Since those equations were the correct ones it is interesting here to see what new features can be inherited in the fluid equations from J in (9.225).

Our first expectation is that no new features will appear when we use (9.225) to derive the fluid equations of change. This is exactly what we find for the equation of continuity. There the property ϕ is m and we have the left side of the equation as

$$m \int d\underline{v}_1 \cdot \left(\frac{\partial}{\partial t} f^{(1)} + \underline{v}_1 \cdot \frac{\partial}{\partial \underline{r}} f^{(1)}\right) = \frac{\partial}{\partial t} \rho + \frac{\partial}{\partial \underline{r}} \cdot \rho \underline{v}_o \qquad (9.226)$$

which is the left side of (8.2). The right side is now found to be

$$- m \int d\underline{v}_1 \int d\underline{r}_2 \int d\underline{v}_2 \left(\underline{a} \cdot \frac{\partial}{\partial \underline{v}_1}\right) f^{(2)} = 0 \qquad (9.227)$$

The zero value comes from (9.224). Thus we do, as expected, find nothing new for the equation of continuity.

To proceed further we need to be able to evaluate other integrals of the type (9.227) where m is replaced by other ϕ_i. For that we need a formula for J which is more explicit.

When we ask for the further development of the integral J (9.222) in the equation for $f^{(1)}$ (9.225) we find ourselves in the middle of the "revolution" in statistical mechanics which occurred in the 1940's. It was not that J was new, for the original Boltzmann equation had the form of (9.225), as mentioned above in (4.4), and J had the form (4.5) of a "collision integral" (appropriate for the two-particle approximation). But the Boltzmann J had been derived from stochastic arguments and not from the two-particle Liouville equation. That dependence on stochastic arguments hid from sight the precise point at which mechanics had to be subsidized by statistics. The revolution in the 1940's leading to the BBGKY theory described in Section 4.4 was aimed at deriving the Boltzmann J from the Liouville equation in order to determine what subsidizing assumptions, if any, were necessary, and then using those to extend the two-particle approximation to the many-particle approximation.

What came out of that revolution were many derivations leading from the Liouville J we have here

$$J_1 = - \int d\underline{r}_2 \int d\underline{v}_2 \left(\underline{a}_1 \cdot \frac{\partial}{\partial \underline{v}_1}\right) f^{(2)} \tag{9.228}$$

to the Boltzmann J_1 (4.5), often expressed as

$$J_1 = 2\pi \int d\underline{v}_2 \int b db \ |\underline{v}_{21}| (f_1' f_2' - f_1 f_2) \tag{9.229}$$

where b is the "impact parameter" and the primes on the f_1 which are $f^{(1)}(\underline{v}_1)$ have the meaning explained in Fig.4.1. The issues and derivations were discussed at the Brussels Symposium in 1956 (Prigogine, 1958). It was not clear at that time what is the exact nature of the subsidy needed from statistical mechanics, and the suggestions offered could not be weighed against each other. There were later attempts to find the "common denominator" (e.g., Hollinger and Curtiss, 1960, 1962), and they did isolate a common denominator, which is known as the "molecular chaos assumption," but it was not clear then how much of this denominator was mechanical and how much was statistical.

We could pursue the development of (9.228) toward (9.229) in order to be able to carry further our discussion about fluid equations of change, and there are many references for that development (e.g., Prigogine, 1958, H. Green, 1952, Kirkwood, 1947, Hirschfelder, 1954, p.449). All of these derivations involve assumptions that mix mechanics and statistics. One assumption, for example, is that, by chance, no third particle interferes during a collision between two particles. If we restrict ourselves to rigid spheres, we escape that problem. We pick up other problems and deficiencies, but they can be dealt with at a later stage.

If we make a commitment to rigid spheres right at the beginning, we find a much simpler path to the exact results we need here. (They can be generalized to cover other cases at a later stage.) The rigid spheres offer an alternative route to our derivations which is simpler. First of all, we can say that the interaction potential u in (9.202) is zero. But we have to compensate for that by saying that $f^{(2)}$ is zero not only at the boundary where the $\underline{r}_1$ become infinite, but also at the inner boundary where the distance $r = |\underline{r}_2 - \underline{r}_1|$ between the two particles is equal to the molecular diameter σ:

$$f^{(2)} = 0, \quad r \le \sigma \tag{9.230}$$

This condition merely makes it explicit that billiard balls cannot invade each other's space. It represents the rigid-sphere potential u without making it necessary to make u explicit.

Using this approach to the integration over $\underline{s}_2$ of the Liouville equation (9.205) with (9.206), in order to obtain (9.225), we find agreement here with (9.219) and (9.223), and we find the integral in (9.222) to be zero because the acceleration $\underline{a}_1$ is zero. But we have to reconsider (9.220) and (9.221) to take into account the inner boundary at $r = \sigma$.

Looking first at (9.220), and let us reduce the problem to one dimension, we have to evaluate

$$\int dx_2 \frac{\partial}{\partial x_1} f^{(2)}(x_1, x_2) \tag{9.231}$$

Previously we could simply interchange the integration and the differentiation, but now we have

$$\left[\int_{-\infty}^{x_1-\sigma} dx_2 + \int_{x_1+\sigma}^{\infty} dx_2 \right] \frac{\partial}{\partial x_1} f^{(2)}(x_1, x_2) \tag{9.232}$$

leaving out the "hole" $x_1 - \sigma < x_2 < x_1 + \sigma$. To evaluate the effect of the hole we need to first evaluate the integral

$$\frac{\partial}{\partial x_1}\left[\int_{-\infty}^{x_1-\sigma} dx_2 + \int_{x_1+\sigma}^{\infty} dx_2\right] f^{(2)}(x_1,x_2) \quad (9.233)$$

Let us review, briefly, the following facts from calculus:

$$\frac{\partial}{\partial x}\int_0^x dy\, f(y) = f(x)$$

$$\frac{\partial}{\partial x}\int_0^x dy\; f(x,y) = f(x,x) + \int_0^x dy\, \frac{\partial}{\partial x} f(x,y) \quad (9.234)$$

Thus we have (9.233) as

$$f^{(2)}(x_1,x_1-\sigma) - f^{(2)}(x_1,x_1+\sigma) + \int_{-\infty}^{\infty} dx_2\, \frac{\partial}{\partial x_1} f^{(2)} \quad (9.235)$$

From that we have

$$\int_{-\infty}^{\infty} dx_2\, \frac{\partial}{\partial x_1} f^{(2)} = \frac{\partial}{\partial x_1}\int_{-\infty}^{\infty} dx_2\, f^{(2)}$$

$$+ f^{(2)}(x_1,x_1-\sigma) - f^{(2)}(x_1,x_1+\sigma) \quad (9.236)$$

Going back to three dimensions requires some manipulations we will omit. For example, instead of taking each direction and moving forward and backward a distance σ, we will sum over all directions $\underline{k}$, where $\underline{k}$ is the unit vector from $\underline{r}_1$ toward $\underline{r}_2$ (see Fig.4.1). More details are offered elsewhere (Hollinger, 1962). We finally obtain

$$\int d\underline{r}_2 \int d\underline{v}_2 \left(\underline{v}_1 \cdot \frac{\partial}{\partial \underline{r}_1}\right) f^{(2)} = \underline{v}_1 \cdot \frac{\partial}{\partial \underline{r}_1} f^{(1)}$$

$$+ \int d\underline{v}_2 \int d\underline{k}(\underline{v}_1 \cdot \underline{k}) f^{(2)}(\underline{r}_1, \underline{v}_1, \underline{r}_1 + \sigma\underline{k}, \underline{v}_2) \qquad (9.237)$$

This gives us (9.220) in the case where we have the "inner" boundary for rigid spheres. Let us look next for (9.221).

The case of (9.221) is similar to that of (9.220). In one dimension we have

$$\int_{-\infty}^{\infty} dx_2 \frac{\partial}{\partial x_2} f^{(2)} = \int_{-\infty}^{x_1-\sigma} dx_2 \frac{\partial}{\partial x_2} f^{(2)} + \int_{x_1+\sigma}^{\infty} dx_2 \frac{\partial}{\partial x_2} f^{(2)} \qquad (9.238)$$

$$= f^{(2)}(x_2 = x_1 - \sigma) - f^{(2)}(x_2 = -\infty)$$

$$+ f^{(2)}(x_2 = \infty) - f^{(2)}(x_2 = x_1 + \sigma) \qquad (9.239)$$

The value of $f^{(2)}$ at infinite x_2 is zero, as assumed before, but the other values cannot be ignored. Summing over all directions by varying the unit vector $\underline{k}$, we have the three-dimensional formula (9.221) as

$$\int d\underline{r}_2 \int d\underline{r}_2\left(\underline{v}_2 \cdot \frac{\partial}{\partial \underline{r}_1}\right) f^{(2)} = \sigma \int d\underline{v}_2$$

$$\cdot \int d\underline{k}(\underline{k} \cdot \underline{v}_2) f^{(2)}(\underline{r}_1, \underline{r}_1 + \sigma\underline{k}) \qquad (9.240)$$

Here we have suppressed from notation the dependence of $f^{(2)}$ on the $\underline{v}_1$ to emphasize the dependence on $\underline{r}_1$, with $\underline{r}_2$ expressed as $\underline{r}_1 + \sigma\underline{k}$. The only $\underline{r}_2$ positions considered are ones where sphere 2 is touching sphere 1 at the direction $\underline{k}$. Those values of $\underline{r}_2$ are, evidently, the only ones that matter in the development of (9.228) by the route we follow.

When we combine the results of the derivations above, we find J in (9.225) as

$$J = \sigma \int d\underline{v}_2 \int d\underline{k}(\underline{v}_{21} \cdot \underline{k}) f^{(2)}(\underline{r}_1, \underline{v}_1, \underline{r}_1 + \sigma\underline{k}, \underline{v}_2) \qquad (9.241)$$

This formula is more interesting if we separate the sum (integration) over directions $\underline{k}$ into the part with $(\underline{k} \cdot \underline{v}_{21}) > 0$ and the part with $(\underline{k} \cdot \underline{v}_{21}) < 0$. From Fig.4.1 we see that the case of $(\underline{k} \cdot \underline{v}_{21}) > 0$ involves $\underline{v}_1$ and $\underline{v}_2$ as post-collision velocities while the case of $(\underline{k} \cdot \underline{v}_{21}) < 0$ involves $\underline{v}_1$ and $\underline{v}_2$ as precollision velocities. The distinction between these cases turns out to be very important, so the separation of (9.241) into the two parts is very desirable. Let us limit the integration over $\underline{k}$ to the $\underline{k}$ (for given $\underline{v}_1$ and $\underline{v}_2$) where $(\underline{k} \cdot \underline{v}_{21})$ is positive. Then we can express (9.241) as

$$J = \pi\sigma^2 \int d\underline{v}_2 \int d\underline{k}\ (\underline{v}_{21} \cdot \underline{k})\Big[f^{(2)}(\underline{r}_1, \underline{r}_1 + \sigma\underline{k}) - f^{(2)}(\underline{r}_1, \underline{r}_1 - \sigma k)\Big] \qquad (9.242)$$

Here, again, we have suppressed the $\underline{v}_i$ from notation in the argument of $f^{(2)}$ to emphasize the importance of the positions $\underline{r}_1$ and $\underline{r}_2 = \underline{r}_1 + \sigma\underline{k}$.

Finally we have arrived at a formula for J which is exact and much less ambiguous than (9.228). But it does still involve $f^{(2)}$ instead of $f^{(1)}$. So we have made little progress toward a replacement for J in (9.225) that would give us the fundamental fluid equations of change on an easy platter. Up to this point we have used no statistics, only mechanics, and we would like to continue this way to derive the fluid equations of change. They should come from mechanics without statistics, as they did in the one-particle approximation (except for the entropy equation). The only clear way to make any progress beyond (9.242) is to somehow express $f^{(2)}$ in terms of $f^{(1)} = f$. Then (9.242) could have a meaning at the fluid level described by one-particle densities and fluxes.

In general we have nowhere to go to find a relation between $f^{(2)}$ and $f^{(1)}$. The only place we know about is equilibrium, where $f^{(2)}$ factors into a product of the $f^{(1)}$ functions (9.127). Let us consider the fluid a short time after a previous equilibrium and notice the Liouville evolution (9.207) together with the factorization (9.127) at that past equilibrium. For $f^{(2)}$ in the second term of (9.242), where the velocities are precollision velocities, going back in time does not change the velocities, and we can factor $f^{(2)}$:

$$f^{(2)}(\underline{r}_1, \underline{r}_1 - \sigma\underline{k}) = f^{(1)}(\underline{r}_1, \underline{v}_1) f^{(1)}(\underline{r}_1 - \sigma\underline{k}, \underline{v}_2) \quad (9.243)$$

But the $f^{(2)}$ in the first term on the right side of (9.242) presents a new challenge. The velocities $\underline{v}_i$ for that $f^{(2)}$ are post-collision velocities, and a collision just occurred (the particles are touching), so we have no reason whatever to assume that the $f^{(2)}$ there is a most probable $f^{(2)}$ or that it factors or that it has any resemblance to the past equilibrium $f^{(2)}$. On the other hand, $f^{(2)}$ cannot change instantaneously, as during the instantaneous collision between rigid spheres. It must have the same value on both sides of the time when that collision occurs. So we have, in this situation,

$$f^{(2)}(\underline{s}_1, \underline{s}_2, t) = f^{(2)}(\underline{s}_1', \underline{s}_2', t) \quad (9.244)$$

where the states $\underline{s}_i$ and $\underline{s}_i'$ have the same positions $\underline{r}_i$ but differ in that the $\underline{v}_i'$ are velocities <u>before</u> a collision that produces the $\underline{v}_i$. Thus the $\underline{v}_i'$ are precollision velocities, and we have, by the argument above, the factorization

$$f^{(2)}(\underline{r}_1, \underline{r}_1 + \sigma\underline{k}) = f^{(1)}(\underline{r}_1, v_1') \; f^{(1)}(\underline{r}_1 + \sigma\underline{k}, \underline{v}_2') \quad (9.245)$$

Let us be clear about the factorizations of $f^{(2)}$ above. We do not assume that the factorization holds in general, or that it is propagated in time or that it is favored in any way except at equilibrium and shortly afterwards. When we think of $f^{(2)}$ evaluated at precollision velocities, we can imagine a recent equilibrium where $f^{(2)}$ factored, or we can use stochastic reasoning to say that precollision velocities are not "correlated" by past collisions, so of course their probabilities factor. That is the "assumption of molecular chaos" in its simplest form, and it works very well, even though it is shallow from our point of view here. What really matters is the recent equilibrium, which not only justifies the practical effectiveness of the assumption of molecular chaos but also explains the psychology behind it in physical terms.

The subtle thing in the argument above is represented by the primes in (9.245). They show that we do not assume that $f^{(2)}$ evaluated at post-collision velocities does factor, but that such an $f^{(2)}$ is equal to an $f^{(2)}$ evaluated at precollision velocities, and that does factor. So the first $f^{(2)}$ can be expressed in terms of a product of $f^{(1)}$ functions, but not $f^{(1)}$ evaluated at the $\underline{v}_i$ which are arguments of $f^{(2)}$. The argument velocities $\underline{v}_i$ have to be replaced by precollision velocities $\underline{v}_i'$.

We now have the collision integral J (9.242) as

$$J = \sigma^2 \int d\underline{r}_2 \int d\underline{k}(\underline{v}_{21} \cdot k)\Big[f_1'(\underline{r})\, f_2'(\underline{r} + \sigma\underline{k}) - f_1(r)\, f_2(\underline{r} - \sigma k)\Big] \tag{9.246}$$

where $\underline{r}$ is the $\underline{r}_1$. This does make the original form (9.228) more explicit, as requested. And it can be used to pursue the fluid equations of change. But (9.246) contains the seed of recent equilibrium, and we did hope to get the equations of change from the Liouville equation without any such appeal to statistics.

Now that we are reminded about the desire to get the equations of change without statistical seeds, let us go back to (9.228), which contains no such seeds, and push harder there to get equations of change. We did find the equation of continuity in (9.226) and (9.227). Let us now look for the equation of motion (the momentum equation) as (8.3) in one dimension or (8.28) in three dimensions, with no external forces. If we multiply (9.225) by $m\underline{v}_1$ and integrate over $\underline{v}_1$, and ignore J (J = 0), we will find our way to (8.28), as we did in the one-particle approximation. If we include J, we add a term to the right side of (8.28) which is

$$\underline{A} = m \int d\underline{v}_1\, \underline{v}_1\, J = -m \int d\underline{v}_1 \int d\underline{v}_2 \int d\underline{r}_2\, \underline{r}_1 \left(\underline{a}_1 \cdot \frac{\partial}{\partial \underline{v}_1}\right) f^{(2)} \tag{9.247}$$

Using the identity

$$\underline{a}_1 \cdot \frac{\partial}{\partial \underline{v}_1} (\underline{r}_1 f^{(2)}) = \underline{v}_1 \left(\underline{a} \cdot \frac{\partial}{\partial \underline{v}_1}\right) f^{(2)} + \underline{a}_1 f^{(2)} \tag{9.248}$$

and imposing the boundary condition (9.224), we find

$$\underline{A} = m \int d\underline{v}_1 \int d\underline{v}_2 \int d\underline{r}_2\, \underline{a}_1\, f^{(2)} \tag{9.249}$$

This looks like some kind of average acceleration of particle 1 due to the forces from other particles. In fact, in view of (9.203), we have

$$\underline{A} = -\overline{\frac{\partial}{\partial \underline{r}}\, nu} = -\int d\underline{v}_1 \int d\underline{v}_2 \int d\underline{r}_2 \frac{\partial u}{\partial \underline{r}_1} \tag{9.250}$$

where $\underline{r}$ is $\underline{r}_1$ and the number density n is a normalizing factor. Thus $\underline{A}$ is the average force on a particle at $\underline{r}$ due to its neighboring particles. This is the term that must be added to the right side of the equation of change (8.28), ignoring the external force,

$$\frac{\partial}{\partial t}(\rho \underline{v}) + \frac{\partial}{\partial \underline{r}} \cdot (\rho \underline{v}\,\underline{v}) = -\frac{\partial}{\partial \underline{r}} \cdot \underline{\underline{p}} + \underline{A} \tag{9.251}$$

It is unfortunate, from our point of view, that $\underline{A}$ turns out to be the average of a gradient rather than the gradient of an average. If $\underline{A}$ were the gradient of an average, such as the one suggested by (9.250), we would have (9.251) in the original form of the fluid equation (8.28) with the pressure tensor modified,

$$\underline{\underline{p}}_{total} = \underline{\underline{p}} + n\bar{u}\,\underline{\underline{U}}$$

$$\underline{\underline{U}} = \begin{pmatrix} 1 & 0 & 0 \\ 0 & 1 & 0 \\ 0 & 0 & 1 \end{pmatrix} \tag{9.252}$$

And that pressure tensor would make a lot of sense. The first part of the pressure tensor is purely kinetic from the molecular point of view, as indicated by (9.118), so it is expected that there should be an addition like the one in (9.252) representing intermolecular forces. But we do have to worry about the difference between the average of a gradient and the gradient of an average. There is no bell ringing here for the status of the form of the fluid equations of change in the two-particle approximation. What is suggested here, and it does work out according to available theories, is that intermolecular forces (or collisions) do not change the form of the fluid equations of change but do change our one-particle formulas for the fluxes of momentum (pressure) and kinetic energy (heat).

We leave the equations of change with some disappointment and approach the equations of flux. It is possible, of course, that the equations of change (8.2) through (8.5) are not as general as we thought before. Maybe there is something wrong with the fluid concept which is corrected only near recent equilibrium. We will not try to answer that. We can say, as we will see below, that current theories about the molecular underpinnings of fluid behavior do suggest that the fluid equations are correct in form but the form of the fluxes varies with the medium under consideration. No general proof of that has been given, and we do not have such a proof, so we have to leave that question up in the air.

As for the equations of flux, they come readily out of the theory outlined above. Let us go back to the original Boltzmann equation (1872), which is (9.225) with J as in (9.246) simplified by ignoring the difference between $\underline{r} = \underline{r}_1$ and $\underline{r} \pm \sigma\underline{k}$ on the right side of (9.246). Thus we have the original Boltzmann equation

$$\frac{\partial}{\partial t} f_1 + \underline{v}_1 \cdot \frac{\partial}{\partial \underline{r}} f_1 = J$$

$$J = \sigma^2 \int d\underline{v}_2 \int d\underline{k}(\underline{v}_{21} \cdot \underline{k})(f_1' f_2' - f_1 f_2) \qquad (9.253)$$

The equations of change implied by (9.253) are the original equations derived from the fluid concept (8.2) to (8.5), and that would be gratifying here if not for the seeds of recent equilibrium so firmly implanted in (9.253).

Let us digress for a moment to the equations of change as implied by the Boltzmann equation (9.253). They are not completely pure or general, but they are worth noticing. We can get the fluid equations of change by multiplying (9.253) through by the corresponding properties $\phi_i(\underline{v}_1)$ and integrating over $\underline{v}_1$. This gives the fluid equations found before with an additional term coming from J which is, for molecular property ϕ,

$$\underline{A} = \sigma^2 \int d\underline{v}_1 \int d\underline{v}_2 \int d\underline{k}(\underline{v}_{21} \cdot \underline{k})\phi_1(f_1'f_2' - f_1f_2) \qquad (9.254)$$

Notice that we can interchange $\underline{v}_1$ and $\underline{v}_2$ (interchange the labels on the particles) without changing the integral (9.254). So we could replace ϕ_1 in the integrand by $1/2(\phi_1 + \phi_2)$ without changing the value of the integral. We could also switch from $\underline{v}_1$ and $\underline{v}_2$ as the integration variables to $\underline{v}_1'$ and $\underline{v}_2'$. The Jacobian of the transformation is unity, so we have

$$d\underline{v}_1' \, d\underline{v}_2' = d\underline{v}_1 \, d\underline{v}_2 \qquad (9.255)$$

which implies (9.254) with ϕ_1 replaced by $-\phi_1'$, or, as seen above, $-1/2(\phi_1' + \phi_2')$. So we have $\underline{A}$ in (9.254) as one-fourth the sum of four equivalent integrals

$$\underline{A} = \frac{\sigma^2}{4} \int d\underline{v}_1 \int d\underline{v}_2 \int d\underline{k}(\underline{v}_{21} \cdot \underline{k}) [\phi_1 + \phi_2 - \phi_1' - \phi_2'] \cdot [f_1'f_2' - f_1f_2] \qquad (9.256)$$

Looking at (9.256) we see that the addition due to $\underline{A}$ to the fluid equations of change is zero for any molecular property ϕ that satisfies

$$\phi_1 + \phi_2 = \phi_1' + \phi_2' \qquad (9.257)$$

which is to say that the sum $\phi_1 + \phi_2$ is conserved in a collision. Such a property is called a "summational invariant." The properties of interest here are summational invariants: mass, momentum, and asymptotic kinetic energy. Thus the form of the equations of change derived from the Boltzmann equation is the same as the form derived from the one-particle Liouville equation.

If we replace the Boltzmann J (9.253) by the Enskog J (9.246) and include the Enskog Y function (Enskog, 1922), we find new terms in the equations of change, but they can all be incorporated in the fluxes for "dense gases" as "collisional contributions" to the fluxes. Thus the original form of the equations of change is ultimately restored.

Continuing to the equations of flux as derived from the Boltzmann equation, we finally reach a subject which reveals the power of that equation (and the Enskog equation) toward applications. In fact we encountered the subject in Chapter 3 where it was described as a major step forward in the development of applications of kinetic theory. Here we only need to review the discussion in Chapter 3.

The Boltzmann equation (3.12) is separated into simpler equations by the perturbation expansion (3.14). The zero-order equation (3.20) can be solved to find the zero-order solution (3.23). Notice that the zero-order solution is the recent equilibrium solution. The peculiar perturbation theory applied to the Boltzmann equation squelched any arbitrary initial conditions and forced the solution, including higher order terms, to be centered around equilibrium. It is not obvious that this equilibrium is playing the role of an initial condition, but in retrospect we see such a role coming into the theory when we try to justify the factorization of $f^{(2)}$ into the one-particle distribution functions on the right side of (3.12). The velocities in the primed f_1 are derived from post-collision velocities $\underline{v}_i$ by going backwards in time to a precollision configuration. That configuration preserves the factoring which occurred naturally at a recent equilibrium.

The zero-order solution having been found as the most probable distribution consistent with local values of $n(\underline{r},t)$, $T(\underline{r},t)$, and $\underline{v}_0(\underline{r},t)$, it can be inserted on the left side of (3.21) to provide the first-order equation for the first correction $f^{(1)}$ to $f^{(0)}$. The derivatives on the left side as applied to (3.23) with v replaced by $(\underline{v}-\underline{v}_0)$, and as modified by equations of change, lead to an expression for the left side as a linear combination of gradients of n, T, and $\underline{v}_0$. The $f^{(1)}$ unknowns in the integrand on the right side of (3.21) with (3.22) inherit the form of the linear combination of gradients, and hence the form of the equations of flux is established even before any attempt is made to solve (3.21). "Solving" (3.21) is a matter of determining how the coefficients of the linear combination depend on velocity $\underline{v}$. When that has been accomplished, then $f^{(1)}$ can be inserted in (3.24) to obtain the equations of flux.

The step of finding the velocity dependence of $f^{(1)}$ is accomplished by expanding the unknown coefficients in powers of $\underline{v}$ or in orthogonal polynomials and solving algebraically to find the first few terms in that expansion (Chapman and Cowling, 1939; Hirschfelder, 1954, Chapter 7).

The results obtained by this scheme are the best we have for predictions of transport coefficients for dilute gases. And they offer not only the "direct" coefficients but also the "cross-over" coefficients in the complex equations of change. The cross-over coefficients do satisfy the Onsager reciprocal relations, which reflects the recent equilibrium.

The justification for the remarkable two-particle approximation to transport phenomena based on the Boltzmann and Enskog equations has been identified as the "molecular chaos" assumption described in Section 4.4. One could say that this assumption is all we have to add to mechanics to obtain the best theory of transport phenomena. But that all is too much! We cannot leave molecular chaos in the air as a stochastic mystery. From our point of view we see the molecular chaos assumption coming from the derivation of (9.126) or (9.127) at a recent past equilibrium.

There is one very important property we have neglected in this section, namely, entropy. In dealing with entropy we have to tread with caution because there are so many candidates for the fluid entropy of thermodynamics but only one correct choice. Let us consider first the Boltzmann H function,

$$H(t) = \int d\underline{r} \int d\underline{v} \; f \ln f \tag{9.258}$$

The time rate of change is given by

$$\frac{dH}{dt} = \int d\underline{r} \int d\underline{v} \; (1 + \ln f) \frac{\partial}{\partial t} f \tag{9.259}$$

The first term on the right side is the rate of change of the total number of particles in the gas, which we can treat as a constant.

Omitting that first term and using the Boltzmann equation to eliminate the time derivative of f, we find (9.259) as

$$\frac{dH}{dt} = \int d\underline{r} \int d\underline{v} \ \ell n f_1 \left[J_1 - \underline{v}_1 \cdot \frac{\partial}{\partial \underline{r}} f_1 \right] \qquad (9.260)$$

The second term on the right integrates to zero because of the boundary condition (9.224). Thus we have (9.260) as

$$\frac{dH}{dt} = \sigma^2 \int d\underline{r} \int d\underline{v}_1 \int d\underline{v}_2 \int d\underline{k} (\underline{v}_{21} \cdot \underline{k}) \ell n f_1 \cdot [f_1' f_2' - f_1 f_2] \qquad (9.261)$$

If we perform the same manipulations here with the $\underline{v}_i$ that were used above to derive (9.256) from (9.254), we find

$$\frac{dH}{dt} = \frac{\sigma^2}{4} \int d\underline{r} \int d\underline{v}_1 \int d\underline{v}_2 \int d\underline{k} (\underline{v}_{21} \cdot \underline{k})$$

$$\cdot \ [\ell n f_1 + \ell n f_2 - \ell n f_1' - \ell n f_2'] \, [f_1' f_2' - f_1 f_2] \qquad (9.262)$$

Boltzmann noticed a most remarkable feature of this formula. If we use

$$A = f_1 f_2 \qquad B = f_1' f_2'$$

then the product of the brackets in (9.262) is

$$\left[\ell n \left(\frac{A}{B}\right)\right] [B - A] \qquad (9.263)$$

This quantity must always be negative. Hence the integrand of (9.262) is negative everywhere, and (9.262) itself must be negative. The function H must decrease in time. For that reason it has been assumed generally that H is negative entropy.

The trouble with H as a candidate for entropy is that H is not defined in terms of the fluid properties $P_i(\underline{r},t)$. There is no reason to believe the f in H is the most probable f consistent with the P_i. The H-theorem described above is important for other reasons, but not as an indicator of the growth of fluid entropy.

One application of the H-theorem is to find the equilibrium f. If we set the time derivative of H equal to zero we can treat (9.262) as an equation to be solved for the equilibrium f. From the implication that $\ell n f$ is a summational invariant, which must be a linear combination of the independent summational invariants (mass, momentum, and kinetic energy), we are led to the familiar formula (9.151).

A second application of the H-theorem is to the zero-order equation (3.20) (modified to make it three-dimensional). If we multiply by $\ell n f_1$ and integrate over $\underline{v}_1$ and $\underline{r}$, we have, again, the equation (9.261) at equilibrium and are led once again to the familiar solution (9.151). Thus the equilibrium f can also be regarded as the zero-order f in the Chapman-Enskog perturbation scheme. (That is no surprise to us.)

Actually we can treat H as an entropy of some kind. If $f^{(1)}$ is treated as the only property of the probability fluid on the one-particle state space for determining the most probable $f^{(2)}$ on the two-particle state space, then the entropy of the probability fluid described by $f^{(1)} = f$ is related to

$$H^{(1)} = \int d\underline{s}_1 \int d\underline{s}_2 \, f^{(2)*} \, \ell n f^{(2)*} \tag{9.264}$$

where the most probable $f^{(2)}$ is given by (9.217). Thus we have

$$H^{(1)} = \int d\underline{s}_1 \int d\underline{s}_2 \, f_1 f_2 (\ell n f_1 + \ell n f_2) \tag{9.265}$$

This is exactly twice the integral (9.258). We can treat negative H as the entropy of the probability fluid described by f, and it must increase monotonically in time for the same reason any statistical entropy must increase (coarse-graining).

The entropy of the fluid in physical space is given by (9.181) if f^* is determined by one-particle properties $P_i(\underline{r}, t)$, as it has been in our models. It does not matter whether we are in the one-particle approximation or the two-particle approximation or any other approximation. And so the form of the equation of change for entropy and the formula (9.195) are not affected by introducing the two-particle approximation.

9.8 Higher Approximations

The BBGKY theory described in Section 4.4 offered an extension of the two-particle approximation of any level of approximation. We are not interested here in the details of that development but we are interested in the role, if any, of recent equilibrium. And if there is such a role, we very much want to ask the general theory to tell us how recent that equilibrium must be. This is a question that has been nagging us for some time, and when we come face to face with the ultimate theory, it is time to raise that question. Other details of the theory are not of interest here (see, for example, Hollinger, 1962, and the references given there).

Unfortunately we cannot get at the general theory without looking at some of the details. Let us start by introducing the Liouville equation for a system of n-interacting particles, with a density of probability $P^{(n)}(\underline{s}^n, t)$ that the system is in state $\underline{s}^n$ at time t,

$$\underline{s}^n = (\underline{s}_1, \underline{s}_2, \ldots, \underline{s}_n) \tag{9.266}$$

$$\underline{s}_i = (\underline{r}_i, \underline{v}_i) \tag{9.267}$$

The Liouville operator L_n is

$$L_n = \sum_{i=1}^{n} \left(\underline{v}_i \cdot \frac{\partial}{\partial \underline{r}_i} + \underline{a}_i \cdot \frac{\partial}{\partial \underline{v}_i} \right) \tag{9.268}$$

and the Liouville equation is

$$\frac{\partial}{\partial t} P^{(n)} + L^n P^{(n)} = 0 \tag{9.269}$$

This equation governs the change of $P^{(n)}$ at any one point $\underline{s}^n$ due to the "flow" of the probability fluid in that very abstract space of 6n coordinates and velocities. It is not yet obvious how that can help us who dwell down in the physical state space and struggle to deal with information coming from the one-particle state space of $\underline{s} = (\underline{r}, \underline{v})$.

The main obstacle against getting information from (9.269) down to the fluid level is the complexity of the $\underline{a}_i$ in (9.268). Each of those accelerations depends on all the other particles

$$\underline{a}_i = \sum_{j \neq i} \underline{a}_{ij} \tag{9.270}$$

$$\underline{a}_{ij} = - \frac{1}{m} \frac{\partial}{\partial \underline{r}_i} u_{ij} \tag{9.271}$$

There is the simplifying feature of symmetry (Newton's third law)

$$\underline{a}_{ij} = - \underline{a}_{ji} \tag{9.272}$$

but that does not help much here. What we need is a separation of the n particles numbered $1, 2, 3, \ldots, n$ into two subsets numbered $1, 2, 3, \ldots, k$ and $k+1$, $k+2$, $\ldots, n$. Then we can distinguish between $\underline{a}_i^{(k)}$, $1 < i < k$, which represents accelerating forces due to the first k particles, from $\underline{a}_i^{(k+1)}$ which includes contributions $\underline{a}_{ij}$ from particles with $j > k$.

With that separation in mind we seek to express L_n as a sum of contributions $-\omega_i$ (the minus sign here for later convenience),

$$L_n = - \sum_{i=1}^{n} \omega_i \tag{9.273}$$

where ω_i depends only on $\underline{s}_1, \underline{s}_2, \ldots, \underline{s}_j$ with $j \leq i$. This can be done by looking at L_n in (9.268) as

$$L_n = \sum_{i=1}^{n} \underline{v}_i \cdot \frac{\partial}{\partial \underline{r}_i} + \sum_{i \neq j}^{n} \underline{a}_{ij} \cdot \frac{\partial}{\partial \underline{v}_i}$$

with (9.272), which allows

$$L_n = \sum_{n=1}^{n} \underline{v}_i \cdot \frac{\partial}{\partial \underline{r}_i} - \sum_{i=1}^{n} \sum_{j=1}^{i-1} \underline{a}_{ij} \cdot \left(\frac{\partial}{\partial \underline{v}_i} - \frac{\partial}{\partial \underline{v}_j} \right) \tag{9.274}$$

Thus we introduce

$$-\omega_i = \underline{v}_i \cdot \frac{\partial}{\partial \underline{r}_i} - \sum_{j=1}^{i-1} \underline{a}_{ij} \cdot \left(\frac{\partial}{\partial \underline{v}_i} - \frac{\partial}{\partial \underline{v}_j}\right) \tag{9.275}$$

With those preparations we can begin to integrate the Liouville equation for n particles down to its reduced form for k particles, involving $P^{(k)}(\underline{s}^k, t)$, where k is less than n. We hope to integrate all the way down to the one-particle level with corrections to the one-particle approximation to provide corrections there coming from all the higher-order considerations. The first integration over $\underline{s}_n$ of (9.269) gives

$$\frac{\partial}{\partial t} P^{(n-1)} + L_{n-1} P^{(n-1)} = J^{(n-1)} \tag{9.276}$$

where

$$J^{n-1} = \int d\underline{s}_n \, \omega_n \, P^{(n)} \tag{9.277}$$

We find similar forms for the lower equations in the hierarchy leading to the general expression

$$\frac{\partial}{\partial t} P^{(k)} + L_k P^{(k)} = J^{(k)} \tag{9.278}$$

$$J^{(k)} = \int d\underline{s}_{k+1} \omega_{k+1} P^{(k+1)} \tag{9.279}$$

This is the hierarchy of equations known as the BBGKY equations. They are merely expressions of Newton's laws for the trajectories of particles, merely reflections of the Liouville equation (9.269), integrated over many times. But they do offer certain conveniences. For example, if we carry the integration all the way to the "bottom," or next to it, we find the $J^{(1)}$ for the two-particle approximation given in (9.228).

The BBGKY hierarchy of equations gives us nothing new over the Newtonian equations for the trajectories of individual particles or for the trajectory of the entire system in the n-particle state-space unless we introduce an alien (to mechanics) element, or assumption. What has been done in the past is to introduce the "molecular chaos" assumption in some form and refer to it as the alien ingredient that is needed to make the mechanics of the BBGKY hierarchy successful. One could, in this context, describe the molecular chaos assumption by saying that $P^{(k+1)}$ factors into $P^{(k)}P^{(1)}$ whenever we have a configuration in which the k+1 particle has a velocity which is precollisional to all future interactions with the other k-particles. Putting that into (9.278) and (9.279) would reduce those equations to a set of k equations for a set of k unknowns with $P^{(1)}$ left as the "pawn" to be played correctly. The details of that scheme are described elsewhere, as was mentioned above.

We are now in a position to start with the Liouville J (9.228) and derive not only the Boltzmann J (9.229) but also corrections to the latter due to interactions and correlations among three and more particles. For example, we can notice three-body effects on $f^{(2)}$ by using (9.278) and (9.279) with k = 2. (We will refer here to $P^{(2)}$ rather than $f^{(2)}$.)

$$\frac{\partial}{\partial t} P^{(2)} + L_2 P^{(2)} = \int d\underline{s}_3 \omega_3 P^{(3)} \tag{9.280}$$

That would be improved further by considering the higher equations (9.278). Notice that each equation for $P^{(k)}$ involves, on the right side, $P^{(k+1)}$, so we cannot treat the equations as "dynamical" or uncoupled unless we go all the way to the top. The top equation for $P^{(n)}$

$$\frac{\partial}{\partial t} P^{(n)} + L_n P^{(n)} = 0 \tag{9.281}$$

is dynamical.

The BBGKY method for extracting information from the set (9.278) of coupled equations is to introduce a perturbation expansion

$$P^{(k)} = P_o^{(k)} + \lambda P_1^{(k)} + \lambda^2 P_2^{(k)} + \ldots \tag{9.282}$$

with λ inserted on the right side of (9.278),

$$\frac{\partial}{\partial t} P^{(k)} + L_k P^{(k)} = \lambda J^{(k)} \tag{9.283}$$

The zero-order equation is

$$\frac{\partial}{\partial t} P_o^{(k)} + L_k P_o^{(k)} = 0 \tag{9.284}$$

and the first-order equation is

$$\frac{\partial}{\partial t} P_1^{(k)} + L_k P_1^{(k)} = \int d\underline{s}_{k+1} \omega_{k+1} P_o^{(k+1)} \tag{9.285}$$

The zero-order equation is a Liouville equation for k particles. There is no coupling to particles outside the group of k particles. Thus the solution is (9.103),

$$P^{(k)}(\underline{s}^k, t) = e^{-tL_k} P^{(k)}(\underline{s}^k, 0) \tag{9.286}$$

We see that the zero-order solution emphasizes initial data. The displacement operator in (9.286) takes the point starting at $\underline{s}_k$ and moves it backwards in time along the k-particle trajectory for time t and then allows $P^{(k)}$ to be evaluated at that new position at the initial time.

To solve the first-order equation (9.285) we see the right side as a known inhomogeneity and the left side as

$$\frac{\partial}{\partial t} e^{-tL_k} P_1^{(k)} = e^{-tL_k}\left[\frac{\partial}{\partial t} P_1^{(k)} + L_k P_1^{(k)}\right] \tag{9.287}$$

Thus we have the solution as

$$P_1^{(k)}(t) = \int d\underline{s}_{k+1} \omega_{k+1} e^{tL_k} e^{-tL_{k+1}} P_o^{(k+1)}(0) \tag{9.288}$$

(Here we are suppressing from notation the $\underline{s}_k$ in the arguments of the $P^{(k)}$.) Thus $P_1^{(k)}$ is found as an integral involving the initial values of $P_0^{(k+1)}$ evaluated at positions $\underline{s}_i'$ derived from the $\underline{s}_i$ by being moved around by the operator in the integrand of (9.288). That operator takes a point at $\underline{s}^{(k+1)}$ and moves it back along the (k+1)-particle trajectory for a time t and then moves the first k particles back in future time t along the k-particle trajectory. If particle (k+1) never interacts with the first k-particles, along the (k+1)-particle trajectory, then they end up where they started.

Let us consider the case of $P^{(1)}$ in first order

$$P_1^{(1)}(\underline{s}_1, t) = \int d\underline{s}_2 \omega_2 e^{tL_1} e^{-tL_2} P_o^{(2)}(\underline{s}_1, \underline{s}_2, 0) \qquad (9.289)$$

A few modifications in the notation will show (9.289) as a familiar quantity. First we replace $P^{(k)}$ by $f^{(k)}$, the distribution function (the different normalizations can be overlooked in this case). Then we refer to the zero-order $P^{(k)}$ as the whole $P^{(k)}$ at the initial time. We also replace the integration variable $\underline{s}_2$ by

$$\underline{s}_2' = e^{-tL_1} \underline{s}_2 \qquad (9.290)$$

which merely moves particle 2 along a one-particle trajectory and does not affect the integral. What we now have is the zero-order $f^{(1)}$ as

$$f_o^{(1)}(\underline{s}_1, t) = e^{-tL_1} f^{(1)}(\underline{s}, 0) \qquad (9.291)$$

and the first-order correction (9.289) as

$$f_1^{(1)}(\underline{s}_1, t) = \int d\underline{s}_2 \omega_2 S_{12} f^{(2)}(\underline{s}_1, \underline{s}_2, 0) \qquad (9.292)$$

where S_{12} is the Bogoliubov (1946) operator

$$S_{12} = e^{tL_1} e^{tL_1} e^{-tL_2} \qquad (9.293)$$

The interesting feature of the scheme described above is the operator S_{12}. Notice that it takes two particles backwards along a trajectory determined by their interaction for a time t and then brings them forward again with no interactions. If they are at precollision velocities to begin with, as in the second term on the right side of (9.246), then S_{12} has no effect. If they are at post-collision velocities, as in the first term on the right of (9.246), then S_{12} changes the velocities $\underline{v}_i$ into precollision velocities $\underline{v}_i'$. If we assume that $f^{(2)}$ factors at the initial time, then, for rigid spheres, the Bogoliubov integral (9.292) is the Enskog J (9.246).

Similar considerations lead to a three-body correction for J involving three particles and successions of collisions along a backwards trajectory to reach the ultimate precollision velocities. In fact, as we work our way up the BBGKY hierarchy of corrections involving effects of many-body interactions, the quantities to be evaluated are all integrals of

$$S_{1,2,3,\ldots,k}\; P^{(k)}(s^k,0) \qquad (9.294)$$

where S displaces $\underline{s}^k$ backwards to a position such that there were no previous collisions among the particles and then "updates" to the present time t by bringing the particles back along one-body trajectories for a time t.

The details of the development of the scheme described above are not of interest here. What is of interest is to find the "end of the road" for mechanics, the precise point at which statistics must enter to provide a deterministic fluid dynamics. The perturbation scheme described above can be carried to completion at all orders (i.e., to "exactness") in the sense of providing a formula for every $f_m^{(k)}$, where k is any number of particles and m is any order. The formulas above zeroth order are integrals like (9.292) with integrands involving various ω_i, S operators, and initial values of the various $f^{(k)}$. (Maybe we should be referring here to the $P^{(k)}$ rather than the $f^{(k)}$, for the two have different normalizations, but that difference will not be important in our discussion.) Up to that point the derivation is strictly Newtonian without the intervention of any alien assumptions. And it goes as far as we can imagine toward fluid dynamics. To breach the gap we have to identify the initial values of the $f^{(k)}$.

The new and alien element introduced at this point in the established BBKGY theory is to take the time t in the S operator to be infinite even though that is not the current time t. In other words, we ignore the initial-value problem and treat the S operators as displacement operators that go "all the way" backwards in time to reaching pre-collision velocities whether or not that time is greater than the time t elapsed from some initial time. Then $f^{(k)}$ can be factored on the stochastic assumption that precollision velocities are not correlated:

$$f^{(k)}(\underline{s}_1, \underline{s}_2, \ldots, \underline{s}_k) = f(\underline{s}_1)\, f(\underline{s}_2) \ldots f(\underline{s}_k) \qquad (9.295)$$

where f is $f^{(1)}$.

That assumption is known as the "assumption of molecular chaos." It is justified by Bogoliubov (1946) and Cohen (1962) and others as a boundary condition at the remote $\underline{s}^k$ reached going back infinitely far in time. The idea is that a "boundary condition" is easier to swallow than a stochastic assumption. Not that there is any reason to expect the factorization at the boundaries of state-space, but neither is there any mechanical objection. After all, boundary conditions are arbitrary. And the factorization is also sometimes treated as an initial condition at remote time, for that, too, is arbitrary.

From the point of view adopted here it is easy to find a justification for the factorization (9.295) which is not stochastic or remote, and which is compelling rather than arbitrary. We refer to the assumption of a recent equilibrium. If the system was at equilibrium at the initial time $t = 0$, then the factorization (9.295) is not a stochastic assumption but a probabilistic necessity which prevails at equilibrium. We are thinking here of a generalization of (9.217).

Identifying recent equilibrium as the cause for the factorization (9.295) offers a number of practical advantages over the more nebulous "causes" mentioned above. In the first place, all such "causes" have possible flaws, but only in the case of recent equilibrium can the flaws be identified, examined, and corrected. Consider, for example, the assumption of molecular chaos. It certainly does not apply to all situations, and how can we decide when it should apply and when it should not? Going to the extreme of the molecular beam, with no collisions, we can see that the intuitive basis for molecular chaos is lost. But for systems less far removed from equilibrium there is no basis for judgment. Even in the case of the dilute gas considered by Boltzmann there was a problem not noticed by Boltzmann but noticed later by Enskog concerning the factorization. What Enskog noticed was that even at equilibrium the factorization holds only for the velocity-dependent parts of the $f^{(k)}$. If the positions $\underline{r}_i$ of the particles are close enough for interactions, then we must include a factor like the Enskog Y function (9.218). Perhaps Enskog was thinking there, in some way, about the importance of recent equilibrium.

Thanks to M.S. Green (1956) we can now applaud Enskog's intuition, and, from our point of view here, applaud Green's intuition. According to Green's "product condition," the particles have to be separated enough from each other that the equilibrium factorization (9.295) does not require some modification. We see that concern as the last problem in the development of nonequilibrium statistical mechanics.

The second advantage we see coming from the assumption of recent equilibrium as the cause for the factorization (9.295), deals with the "divergence problem" of modern extensions based on the BBGKY theory (Brush, 1972). According to the BBGKY theory based on the molecular chaos assumption, the factorization (9.295) depends on the time t in the S operator being large enough to go back through collisions far enough to reach precollision velocities for k interacting particles. For two touching rigid spheres, this time is negligible, but for three rigid spheres, even if two are touching to start with, the time to go back and get behind their collisions can be considerable. As we move to more particles and softer collisions, the time t to get behind the collisions can grow drastically, and it does, leading to integrals in the perturbation series that do not shrink fast enough to allow the series (9.282) to converge.

From our point of view, based on prejudice toward considerations of a recent equilibrium, we do not have to go back in time further than the finite time required to reach that recent equilibrium. The factorization occurs there whether or not a pursuit of the particles back further in time leads to collisions. So we have no "convergence problem."

One question that has nagged us from the beginning of our considerations about recent equilibrium is: How recent is recent? From the point of view of the BBGKY theory with our assumptions about recent equilibrium added to the mechanics, we can begin to give an answer to that question. First, we may say, that recent equilibrium does not have to be very recent, as it seemed to in the previous sections. It only has to be back there somewhere. But there are practical considerations for theoretical developments. The further back that recent equilibrium is, the greater time allowed for higher order collisions and correlations. We have to include more corrections of higher order. At the same time, the burden of those higher order corrections also depends on the density of the gas and on the range of the interactions. There are many factors involved, and it is not yet understood what roles they play.

There is also the experience we have with fluid dynamics. Many equations which are derivable only at early stages in the withdrawal from equilibrium seem to hold up (their form does not change) for the extended time to reach the next equilibrium. On the other hand, we have to make hard distinctions between solids, liquids, and gases, and those are not explicit in the theory as described above. Needless to say, the theoretical situation is not clear. But it does seem to us that the role of recent equilibrium is important.

CHAPTER 10

IRREVERSIBILITY IN QUANTUM STATISTICAL MECHANICS

Quantum mechanics has its own probabilities to be added to, or identified with, the probabilities of statistical mechanics. We ignore the "identified with" and regard the two kinds of probabilities as different as day and night.

Quantum effects usually add to fluid dynamics nothing more than recognizing the coupling of atoms into stable molecules, which is chemistry. So if we recognize the chemical compositions of fluids, we have already incorporated the quantum effects and can then go back to classical fluid dynamics to follow the rest of the evolutions. That is why we do not have courses on quantum fluid dynamics. But if we do have a need to recognize quantum effects other than chemistry, then we do have to start with the Schrödinger equation or its equivalent and find something new that is not prescribed by classical mechanics.

The Schrödinger equation itself is as deterministic as any other equation of physics, we assume, and it agrees in many respects with Newton's equation. If we want to predict particle motion as such, then we can reduce Schrödinger's equation to Newton's without the interference of uncertainty principles or ideas about how measurements destroy data, as if that were beyond the scope of Newtonian physics. But if we focus on phenomena that can be described by Schrödinger and not by Newton (e.g., the standing wave phenomena that dominate chemistry), then we do encounter radically new models, symmetry principles, etc. Especially in the cases of many-body systems that are cold or dense, we find ourselves in a new world.

Thus we have the question: Suppose we do not seek chemical phenomena, low temperature phenomena, etc., but merely seek to describe phenomena that can be described very well, in principle, by classical statistical mechanics. Is there an advantage in the quantum description? Is it simpler? Does it go farther? Does it expose more clearly the origin of irreversibility?

The first question for us, if we are trying to compare classical and quantum statistical mechanics, is this: Granted that the Schrödinger equation is some kind of generalization of Newton's equation, what is the analog that generalizes the Liouville equation? The answer seems to be in the direction of the unpublished but celebrated Pauli equation, or the Schrödinger equation under "second quantization," or the "diagram techniques," or "field theoretic methods," all devoted especially to perturbation expansions.

If we look for a simple way to describe the expansion of a gas of noninteracting particles we find a hole in quantum theory. It is not easy to get "down" to the level of the required description. So we have problems ahead if we want to reduce quantum models to the point where irreversibility can be the issue of main interest.

What we will do is consider the Schrödinger equation and its variety of forms and then develop one-body and two-body approximations for the use of the Schrödinger equation. We would like to make these developments parallel to the developments of the previous chapter on classical mechanics. We think this can be done but we have not done it, especially in the area of the one-body approximations. The literature for quantum mechanics is dominated in that approximation by considerations of the effects of external forces, especially electromagnetic forces, and never the case of no forces with a chance to expand. So we aim especially for the section on two-body approximations and use the earlier discussions as an introduction to that. In the section on the two-body approximation we do encounter the question of how irreversibility can enter the description. What we find there is the same reference to recent equilibria that we found above in classical mechanics.

10.1 The Schrödinger Equation

We consider a collection of n identical particles (bosons) each of mass m with coordinates $\underline{q} = q_1, q_2, q_3, \ldots.$ for which the wavefunction is

$$\psi(\underline{q}, t) = \psi(q_1, q_2, \ldots q_{3n}, t) \tag{10.1}$$

This function obeys the time-dependent Schrödinger equation,

$$i\hbar \frac{\partial}{\partial t} \psi = H\psi \tag{10.2}$$

where H is the Hamiltonian operator for the system. The eigenfunctions $\psi_n(\underline{q})$ of H satisfy

$$H\psi_n = E_n\psi_n \tag{10.3}$$

where the E_n are the energy levels of the stationary states. We will not give explicit attention to possibilities of degeneracy.

At any time t, ψ belongs to a set of functions, any of which can be expanded in the ψ_n,

$$\psi(\underline{q},t) = \sum_n c_n(t)\psi_n(\underline{q}) \tag{10.4}$$

and hence the time dependence of ψ can be described in terms of the time dependence of the coefficients $c_n(t)$. The ψ_n can be selected so that they are normalized and orthogonal

$$\int \psi_n^*\psi_m d\underline{q} = \begin{cases} 1, & n=m \\ 0, & n\neq m \end{cases} \tag{10.5}$$

and hence we have a way to find the initial values of the c_n from $\psi(\underline{q},0)$:

$$\int \psi_n^*(\underline{q})\,\psi(\underline{q},0)\,d\underline{q} = \sum_m c_m(0) \int \psi_n^*(\underline{q})\,\psi_m(\underline{q})\,d\underline{q} = c_n(0) \tag{10.6}$$

The formulas above apply to systems composed of one particle or any number of particles. Newtonian mechanics has similar generalized formalisms but there is a difference. Newtonian schemes for many particles generally involve sets of equations, not just one, until we reach the Liouville equation. (The Liouville equation is equivalent to $F = ma$ but the form is very different.)

One way to modify the Schrödinger equation to bring it closer to applications to many particle systems is to introduce at the outset a set of properly symmetrized eigenfunctions. Suppose we have a system with no interactions among the particles. The "particles" here could be normal modes or any "independent" or "uncoupled" subsets of coordinates. Physicists like to give names to the "quasi-particles," which may be actual particles. A freely translating particle could be a "translation," or better, a "moton;" a normal mode of vibration is a "vibron" and one quantum's worth is a "phonon;" rotational coordinates represent a "roton;" and we also have "photons, migrons, spinons," etc. But whatever we call them, their coordinates are uncoupled. We have in mind an ideal gas of motons, but we will refer to "ons."

The wavefunction for two ons can be factored in a stationary state,

$$\psi(q_1,q_2) = A(q_1)B(q_2) \tag{10.7}$$

where A and B are one-on-wavefunctions. This factored wavefunction can satisfy the Schrödinger equation and the boundary conditions. But there is an additional requirement. For bosons, for example, we may say that the wavefunction must be symmetric with respect to an interchange of the coordinates

$$\psi(q_1,q_2) = \psi(q_2,q_1) \tag{10.8}$$

and this is not satisfied by (10.7). So we have to replace (10.7) by its symmetrized improvement

$$\psi(q_1,q_2) = A(q_1)B(q_2) + A(q_2)B(q_1) \tag{10.9}$$

Strictly speaking, if we want to maintain normalization in the sense of (10.5), assuming A and B are separately normalized, we have to multiply the right side of (10.9) by $2^{-1/2}$. In the case of $\psi = AA$ or $\psi = BB$ then the wavefunction is already symmetric and we do not have to modify it in any way.

If we have not two ons but many ons, say n ons, then the symmetrization procedure is more complicated. First of all we need to recognize the "distribution" $\underline{n} = (n_1, n_2, n_3, \ldots)$ of ons into one-on states where n_1 is the number in state 1, n_2 is the number in state 2, etc. This is the same distribution that Boltzmann had to consider to describe the way particles are distributed over classical states. In fact we will need the famous Boltzmann formula for the number of ways of assigning particles to the different states,

$$W(\underline{n}) = \frac{n!}{n_1!\, n_2!\, n_3! \ldots} \tag{10.10}$$

To "construct" the properly symmetrized wavefunction for a stationary state $\psi(\underline{n})$ we start out by writing one possible term

$$\psi_o(\underline{n}) = A_1(1)A_1(2)A_1(3) \ldots A_1(n_1)A_2(n_1+1) \ldots . \cdot A_2(n_1+n_2)A_3(n_1+n_2+1) \ldots . \qquad (10.11)$$

where A_1, A_2, ... are the one-on wavefunctions and the n_i keep track of how many particles are assigned to each one-on state. Then we form other terms by permuting the one-ons in such a way as to put them in different states. Finally we have our function $\psi(\underline{n})$ as

$$\psi(\underline{n}) = \frac{1}{\sqrt{W}} \sum_P P\psi_o \quad \text{[above in (10.11)]} \qquad (10.12)$$

where the symbol P refers to a permutation of the type described above, and W is the total number of terms in the sum, i.e., the total number of ways to assign the ons to different states while maintaining the distribution $\underline{n}$. The W factor maintains normalization, which is essential in order to simplify later counting problems. And that factor is the difficult part of the symmetrization procedure. Fortunately we recognize W as the Boltzmann W in (10.10).

At this point we can replace (10.4) by

$$\psi(\underline{q},t) = \sum_{\underline{n}} c(\underline{n},t)\psi(\underline{n}) \qquad (10.13)$$

which makes more explicit the use of the one-on wavefunctions and the distributions $\underline{n}$. It is important to notice that the $\psi(\underline{n})$ are still functions of $\underline{q}$. We have suppressed $\underline{q}$ from the notation to emphasize the organization into terms of various $\underline{n}$.

From here we can go almost directly to a form of the Schrödinger equation that comes very close to our classically oriented requirements. We need only introduce an operator η_i defined by

$$\eta_i \psi(\underline{n}) = n_i(\underline{n}) \psi(\underline{n}) \qquad (10.14)$$

The eigenvalue $n_i(\underline{n})$ is the number of particles in state i if the total distribution is $\underline{n} = (n_1, n_2, \ldots, n_i, \ldots)$, and so the operator η_i merely probes into $\underline{n}$ to find n_i. This sounds very arbitrary as a way of defining an operator, but we can define an operator by specifying its eigenvalues when it is applied to all eigenfunctions in a complete set. In fact, this operator is well known as the "occupation number" operator, often expressed as a product of "creation" and "destruction" operators. If we were going to develop applications in detail and face the problems of counting up terms which arise in that work, we would digress to consider the creation and destruction operators and that whole formalism of "second quantization" (very poorly named), but we do not need such aids to reach the goal of identifying the cracks where irreversibility can creep into quantum mechanics.

The advantage offered by (10.14) can be illustrated readily as follows. Suppose we have the evolved wavefunction $\psi(\underline{q}, t)$ which contains all possible information about the system at time t. And suppose that what we want to know are the populations $n_i(t)$. Since we have an operator for n_i we can use the quantum postulate

$$n_i(t) = \int \psi^*(q,t)\, \eta_i \psi(q,t)\, dq \qquad (10.15)$$

The $n_i(t)$ obtained this way is sometimes referred to as the "average" value or "expectation" value, but whatever its name it is the "predicted" value based on the Schrödinger equation. In view of (10.13) we have $n_i(t)$ as

$$n_i(t) = \int \Big[\sum_{\underline{n}} c^*(\underline{n},t)\, \psi^*(\underline{n}) \Big] \sum_{\underline{n}'} n_i(\underline{n}')\, c(\underline{n}',t)\, \psi(\underline{n}')\, dq$$

$$= \sum_{\underline{n}} \sum_{\underline{n}'} c^*(\underline{n},t)\, c(\underline{n}',t)\, n_i(\underline{n}') \int \psi^*(\underline{n})\, \psi(\underline{n}')\, dq \qquad (10.16)$$

In view of the orthogonality of the $\psi(\underline{n})$ and their normalization, this is

$$n_i(t) = \sum_{\underline{n}} |c(\underline{n},t)|^2 n_i(\underline{n}) \qquad (10.17)$$

Thus the problem of predicting the evolution of the distribution $\underline{n}$ reduces to that of predicting the evolution of $c(\underline{n},t)$.

The focus on evolving $c(\underline{n},t)$ can be accomplished by substituting (10.4) and (10.3) into (10.2)

$$\sum_{\underline{n}} i\hbar\, \dot{c}(\underline{n},t)\psi(\underline{n}) = \sum_{\underline{n}} c(\underline{n},t)E(\underline{n})\psi(\underline{n}) \qquad (10.18)$$

Multiplying both sides by $\psi^*(\underline{m})$ and integrating over $\underline{q}$, using the orthogonality of the $\psi(\underline{k})$, we find

$$i\hbar \frac{d}{dt} c(\underline{m},t) = E(\underline{m})c(\underline{m},t) \qquad (10.19)$$

$$c(\underline{m},t) = c(\underline{m},0)e^{-\frac{iE(\underline{m})t}{\hbar}} \qquad (10.20)$$

We now have everything needed to calculate $\psi(q,t)$ or $c_n(t)$ or, in the "occupation-number representation," $c(\underline{n},t)$ or $n_i(t)$. The initial values for the $c(\underline{n},t)$ can be found from $\psi(q,0)$ in a manner similar to that used in (10.6) to find $c(\underline{n},0)$,

$$c(\underline{n},t) = \int \psi_n^*(\underline{q})\psi(\underline{q},0)\, dq \qquad (10.21)$$

It would appear that we can fulfill the dreams of Boltzmann and his critics by following the evolution of $\underline{n}$ without any appeal to statistics. Actually, we can do just that. But it has not been tried for various reasons. Perhaps the main reason is that we do not know what to use for the initial value $\psi(q,0)$.

The approaches toward predictions of time evolution which have been developed extensively in the literature go in another direction. Rather than assume we know the eigenfunctions of H and can carry out the program outlined above starting with various initial conditions, it is generally assumed that we know how to deal with initial conditions and have the eigenfunctions for an H that is simpler than the H which prevails but wonder how the initial conditions (stationary under the simpler H) evolve when we apply the more complicated H. Let us separate H into two parts

$$H = H_o + \lambda H_1 \tag{10.22}$$

where H_0 is well understood with known eigenfunctions, H_1 is a new perturbing influence, and λ is a parameter ("perturbation parameter") that will eventually be evaluated at $\lambda = 1$ to recover the original problem. We expand the "solution"

$$\psi(\underline{q}, t) = \psi^{(0)}(\underline{q}, t) + \lambda\psi^{(1)}(\underline{q}, t) + \lambda^2\psi^{(2)}(\underline{q}, t) + \ldots . \tag{10.23}$$

and search for each separate term, hoping to quit after two terms or hoping to throw a lot of stuff away and find an infinite-order perturbation expansion.

For example, we could assume that H_0 is the Hamiltonian for a system of noninteracting one-ons and H_1 is an interaction potential energy. Then we would be following the effects of the interactions on the evolution of the system. That is a most appealing goal, and it becomes all the more appealing when we realize that all we are after is a macroscopic rate equation at the initial time, from which we can go to the macroscopic realm to follow the evolution. The perturbation theory is especially appropriate because the need for higher order terms decreases as the time itself backs up toward the initial time. At least that is the case in most time-dependent perturbation theories. But something strange comes up in quantum theory. Or else a strange twist has been forced into the theory. Rather than look for an initial rate equation, many workers have looked for the rate equation that evolves ultimately as the system approaches an equilibrium at large times. It is not at all clear what is the goal in those treatments of time-dependent perturbation theory. We would like to identify the purpose as the formulation of rate equations at the initial time for systems leaving equilibrium, but that may be short sighted.

If we choose to go the route of perturbation theory, which has many advantages in problems of practical interest, then we substitute (10.22) for H in (10.2) and express (10.3) as

$$H_o \psi_n = E_n \psi_n \tag{10.24}$$

We admit that what we know are the energy levels and eigenfunctions for H_0 rather than the full H. We also modify (10.4) to

$$\psi(\underline{q}, t) = \sum_n c_n(t)\, e^{-\frac{iE_n t}{\hbar}} \psi_n(\underline{q}) \tag{10.25}$$

in which we have pulled out of $c_n(t)$, and made explicit, the time dependence that would prevail without the perturbation. This reduces the burden on $c_n(t)$. If there were no perturbation we would find those c_n to be constant and here we ask for the change in such c_n due to the perturbation.

If we substitute (10.25) into the Schrödinger equation (10.2) and multiply both sides on the left by ψ_m^* and integrate to use the orthogonality of the ψ_k, we find

$$i\hbar \dot{c}_m e^{-\frac{iE_m t}{\hbar}} = \lambda \sum_n c_n H_{mn} e^{-\frac{iE_n t}{\hbar}} \tag{10.26}$$

where

$$H_{mn} = \int \psi_m^* H_1 \psi_n \, dq \tag{10.27}$$

Introducing

$$\omega_{mn} = \frac{E_m - E_n}{\hbar}$$

we can express (10.26) as

$$\dot{c}_m = \frac{\lambda}{i\hbar} \sum_n c_n H_{mn} e^{i\omega_{mn} t} \tag{10.29}$$

At this point we expand the c_n in λ, after the manner of (10.23), and we have

$$c_m = c_m^{(0)} + \lambda c_m^{(1)} + \lambda^2 c_m^{(2)} + \ldots \tag{10.30}$$

$$\dot{c}_m^{(r)} = \frac{1}{i\hbar} \sum_n c_n^{(r-1)} H_{mn} e^{i\omega_{mn} t} \tag{10.31}$$

Using the initial values

$$c_m^{(0)} = c_m(0) \qquad c_m^{(r)}(0) = 0, \; r > 0 \tag{10.32}$$

we have a scheme for successively calculating higher and higher order $c_n^{(r)}$.

The execution of this prescription leads to the famous (but unreferenced) Pauli Equation, involving successively higher terms with more time-integrations, abbreviated sometimes in terms of the "time-development operator" $U(\underline{q},t)$,

$$i\hbar \frac{\partial}{\partial t} \psi = H\psi \qquad \psi(\underline{q},t) = U(t)\psi(\underline{q},0) \tag{10.33}$$

where U is, abbreviated,

$$U = e^{-\frac{tH}{i\hbar}} \tag{10.34}$$

Alternatively, we can always move from $c_n(t)$ to the "density matrix" $\rho_{nm} = c_n^* c_m$ with the equation (10.29) presented as

$$\frac{d}{dt} \rho_{nm} = \dot{c}_n^* c_m + c_n^* \dot{c}_m \tag{10.35}$$

which is

$$\frac{d}{dt} \rho_{nm} = \frac{1}{i\hbar} \sum_k \left[H_{kn} e^{i\omega_{kn} t} \rho_{km} - H_{mk} e^{i\omega_{mk} t} \rho_{nk} \right] \tag{10.36}$$

and can develop that perturbation scheme successively. To simplify the algebra involved we can go for the "diagram techniques" or "second quantization" or try to show that many terms are eventually suppressed while the predominant terms (for decay) prevail in the approach to the next (remote) equilibrium.

To simplify the algebra we can go for a lot of things that have become useful techniques, but we never get away from certain elementary obstacles. For example, if the perturbing Hamiltonian does not depend explicitly on time, as in all cases of interest here, then the exact form of (10.25) is

$$\psi(\underline{q}, t) = \sum_n c_n(0) e^{-\frac{iE_n' t}{\hbar}} \psi_n'(\underline{q}) \qquad (10.37)$$

which is a superposition of sinusoidal functions (time-variation). Here E_n' and ψ_n' are eigenvalues and eigenfunctions of the exact Hamiltonian which includes the perturbation. Thus the exact "solution" can be presented as a linear combination of sinusoidal functions. So how can approximate solutions deviate from that and still be useful? The answer is that the exact solution may be a Fourier transform of a solution that decays to equilibrium and then eventually leaves that equilibrium, periodically, to repeat the decay and withdrawal, etc. If this is the case, it would be very useful to have an approximate description of the decay whether or not that description can extend to ends of decay plateaus, and that should be the goal of quantum descriptions of rates just as it is the goal of classical descriptions. Once again we see that the object of the mechanical approach is to describe the forced withdrawal from equilibrium rather than the fluctuational events in the later life of a mechanical system.

10.2 The One-Particle Approximation

Given the Schrödinger equation and a new external force imposed at the initial time, we can find the evolution of the system (one particle) from an initial condition. The simplest example, perhaps, is the case of a particle initially in a stationary state in a one-dimensional box of length L_1 suddenly allowed to expand into a larger box of length L_2. If we follow the density P of probability that the particle is on some interval $a < x < b$

$$P(t) = \int_a^b dx \, \psi^* \psi \qquad (10.38)$$

we find strong oscillations, of course, but in the middle of a period the oscillations are subdued in a way that suggests short equilibrium plateaus.

It would be interesting to follow the one-particle P for the case of many noninteracting particles distributed initially over the stationary states for the original box. If the symmetry requirements can be ignored, the evolution of P can be found easily and there are very long equilibrium plateaus, as in classical mechanics. The derivation of P(t) for the case of bosons or fermions, however, is surprisingly difficult and will not be attempted here. It adds nothing new to our collection of clues concerning the origin of irreversibility.

The simplest one-particle approximation developed in textbooks is the case of a particle in a stationary state under a Hamiltonian H_0 suddenly perturbed by an additional force represented by H_1 and hence forced to leave the initial state. This is the model for spectroscopy, for example, if we take H_1 to be perturbing electromagnetic forces which vary with time. Let us restrict ourselves here to a perturbation H_1 that is not time-dependent. Then we have (10.33) in the form of (10.36) with no time dependence in the H_{ij}.

To be specific, suppose we have a particle initially in state ψ_i and then turn on the perturbation H_1. Initially the density matrix of (10.36) is

$$\rho_{ii}^{(0)} = 1 \quad \rho_{nm}^{(0)} = 0 \quad \text{if } n \neq i \text{ and/or } m \neq i \tag{10.39}$$

Multiplying the right side of (10.36) by λ and then expanding the density matrix in λ we have the hierarchy of successive approximations for (10.36) as

$$\rho_{nm} = \rho_{nm}^{(0)} + \lambda\rho_{nm}^{(1)} + \lambda^2\rho_{nm}^{(2)} + \ldots.$$

$$\frac{d}{dt}\rho_{nm}^{(r+1)} = \frac{1}{i\hbar}\sum_k \left[H_{kn}e^{i\omega_{kn}t}\rho_{km}^{(r)} - H_{mk}e^{i\omega_{mk}t}\rho_{nk}^{(r)}\right] \tag{10.40}$$

Thus the zero-order $\rho_{nm}^{(0)}$ are constant and we can ask those $\rho_{nm}^{(0)}$ to absorb the initial data (10.39),

$$\rho_{ii}^{(0)} = 1, \text{ other } \rho_{ij}^{(0)} = 0 \tag{10.41}$$

$$\frac{d}{dt}\rho^{(1)}_{nm} = \frac{1}{i\hbar}\sum_k \left[H_{kn}e^{i\omega_{kn}t}\delta_{ki}\delta_{mi} - H_{nk}e^{i\omega_{mk}t}\delta_{ni}\delta_{ki}\right] \quad (10.42)$$

where $\delta_{ij} = 0$ unless $i = j$. We see that the derivative of $\rho^{(1)}_{nm}$, and hence $\rho^{(1)}_{nm}$ itself, is zero unless $n = i$ or $m = i$ and $m \neq n$. Thus we have, for $n \neq i$,

$$\frac{d}{dt}\rho^{(1)}_{ni} = \frac{1}{i\hbar}H_{in}e^{i\omega_{in}t}$$

$$\frac{d}{dt}\rho^{(1)}_{in} = -\frac{1}{i\hbar}H_{ni}e^{i\omega_{ni}t} \quad (10.43)$$

with integrals

$$\rho^{(1)}_{ni} = -\frac{H_{in}}{\hbar\omega_{in}}\left(e^{i\omega_{in}t} - 1\right)$$

$$\rho^{(1)}_{in} = \frac{H_{ni}}{\hbar\omega_{ni}}\left(e^{i\omega_{ni}t} - 1\right) \quad (10.44)$$

We notice that the diagonal element ρ_{nn} in first order is zero, and we have learned nothing at that level about the evolution of the ρ_{nm}. But we will get what we want in second order.

In second order we have (10.36) as

$$\frac{d}{dt}\rho^{(2)}_{nm} = \frac{1}{i\hbar}\sum_k \left[H_{kn}e^{i\omega_{kn}t}\rho^{(1)}_{km} - H_{mk}e^{i\omega_{mk}t}\rho^{(1)}_{nk}\right] \quad (10.45)$$

At this point we can simplify the algebra by admitting that we are not going beyond second order and hence the only second order ρ_{nm} of interest is the diagonal element ρ_{nn}. So we rewrite (10.45) with $n = m$ and insert the appropriate formulas for the first order ρ_{nm}, and get, for $n \neq i$,

$$\frac{d}{dt}\rho^{(2)}_{nn} = \frac{2H_{in}H_{ni}}{\hbar^2\omega_{ni}}\sin(\omega_{ni}t) \quad (10.46)$$

$$\frac{d}{dt}\,\rho_{ii}^{(2)} = -\,\frac{2}{\hbar^2}\sum_n{}' \frac{H_{ih}H_{ni}}{\omega_{ni}}\,\sin(\omega_{ni}t) \qquad (10.47)$$

It is generally acknowledged that the levels E_n cannot be identified with precision in a time-dependent description so we should replace each ω value by a continuum of values and replace the sine function in (10.47) by

$$\int_{-\infty}^{\infty} d\omega\; d(\omega)\,\frac{\sin\,\omega t}{\omega} \qquad (10.48)$$

where the density of levels $d(\omega)$ can be treated as a constant $d(0)$ to obtain (10.48) as

$$d(0)\int_{-\infty}^{\infty} \frac{\sin\,\omega t}{\omega}\,d\omega = d(0)\,\pi\,\frac{t}{|t|} \qquad (10.49)$$

The possibility of ignoring ω-dependence in the H_{in} and $d(\omega)$ is due to the fact that the integrand of (10.49) has a steep maximum at $\omega = 0$. Whether or not this procedure can be justified, it is the standard procedure for the present context and it is the backbone of all spectroscopic tests of the time-dependent Schrödinger equation.

According to the procedure outlined above, we have (10.46) presented as

$$\frac{d}{dt}\,\rho_{nn} = T_{in} \qquad (10.50)$$

where

$$T_{in} = \frac{2}{\hbar^2}\,|H_{in}|^2\,\pi\,\frac{t}{|t|} \qquad (10.51)$$

is the "transition probability" per unit time for a transition from the initial state i to another state n. The factor $t/|t|$ in (10.49) and (10.51) is often omitted, even though it is explicit in better tables of integrals, and for good reason. (Many authors have not been interested in questions about time reversibility.) The explicitness of that factor here shows that transition probabilities change sign under time reversal, just as do rate constants, transport coefficients, and other constitutive parameters in rate equations.

If we repeat the derivation above for a more general initial condition (ρ_{nm} initially diagonal) than (10.41), we find, in the place of (10.50),

$$\frac{d}{dt}\rho_{nn} = \sum_{k\neq n} T_{kn}\rho_{kk} - \sum_{k\neq n} T_{nk}\rho_{nn} \tag{10.52}$$

which is supposed to hold (to second order) at the initial time, and we hope, at later times, If we want the rate equation itself at the initial time (time is small) then it is not necessary that the perturbation also be small. This is not to say that a rate equation evaluated at the initial time must retain its form at the macroscopic level into later times, but experience tells us that macroscopic rate equations do retain form long enough to reach the next equilibrium in most cases. Proving that would require an extension of the perturbation expansion to higher order, and that is beyond our scope.

The form of (10.52) is the general form of "master" equations, "rate" equations, i.e., "stochastic" equations that have been suggested for the process under consideration. It is true that the equation is reversible in time, contrary to popular classifications, and it is faithful to the underlying particle dynamics at the initial time and possibly for some short time thereafter. But the faithfulness certainly fades in time and it is no surprise that (10.52) does not have periodic solutions. Long-range properties including periodicity were certainly killed off in the derivation. If this makes the equation "dissipative," let it be so, but let us not construe the dissipation to imply a mechanical denial of the periodicity that was lost due to the fading reliability of the equation or the time reversibility that is a universal characteristic of all equations in our context.

A more explicit way to extend the one-particle approximation to the many-body problem with an external force is offered by the occupation number representation, also referred to mysteriously as "second quantization" or "field theoretic." For our limited purposes we can observe this approach without introducing explicitly the "creation" and "destruction" operators, which help with counting problems, or the "diagrams" or "resolvent sets" etc. All we want is to see where irreversibility enters.

From (10.17) we have

$$n_i(t) = \sum_{\underline{n}} \rho(\underline{n}, \underline{n}, t)\, n_i(\underline{n}) \tag{10.53}$$

where $\underline{n} = (n_1, n_2, \ldots)$ is a distribution of particles over one-particle states and ρ is the density matrix

$$\rho(\underline{n}, \underline{n}, t) = c^*(\underline{n}, t)\, c(\underline{n}, t) \tag{10.54}$$

Let us assume that the initial distribution is $\underline{n}_0$. Then we have the master equation (10.52) as

$$\frac{d}{dt} \rho(\underline{n}, \underline{n}) = \sum \left[T(\underline{k}, \underline{n})\, \delta(k, \underline{n}_0) - T(\underline{n}, \underline{k})\, \delta(\underline{n}, \underline{n}_0) \right] \tag{10.55}$$

where $\delta(\underline{n}, \underline{m})$ is unity if $\underline{n} = \underline{m}$ and zero otherwise. Thus the rate of change of n_i in (10.53) can be put into the form

$$\frac{d}{dt} n_i = \sum_{\underline{n}} T(\underline{n}_0, \underline{n}) [n_i(\underline{n}) - n_i(\underline{n}_0)] \tag{10.56}$$

where the transition probability is

$$T(\underline{n}_0, \underline{n}) = \frac{2}{\hbar^2} |H(\underline{n}_0, \underline{n})|^2 \frac{\sin \omega(\underline{n}_0, \underline{n}) t}{\omega(\underline{n}_0, \underline{n})} \tag{10.57}$$

The notation here is analogous to that of (10.51) but we elevate a subscript n to an argument $\underline{n}$ to be reminded that $\underline{n}$ is not just an intéger but a set of integers. We have reverted to (10.46) in our analogy and have not yet invoked the integration of (10.49). We could just as well invoke that integration here, but we need (10.57) as it stands for a later reference.

The integral $H(\underline{n}_o, \underline{n})$ is

$$H(\underline{n}_o, n) = \int \psi^*(\underline{n}_o) H_1 \psi(\underline{n})\, dq \tag{10.58}$$

with

$$dq = d\underline{r}_1 d\underline{r}_2 \ldots$$

$$d\underline{r}_i = dx_i dy_i dz_i$$

The potential energy H_1 for the perturbing (external) force is a sum of terms for the separate particles,

$$H_1 = \sum_k h(\underline{r}_k) \tag{10.59}$$

Using the boson ψ functions defined by (10.12) we have the term of (10.58) involving $h(\underline{r}_j)$ as

$$\sum_P \sum_{P'} \frac{1}{\sqrt{W(\underline{n}_o) W(\underline{k})}} \int dq\, PA_1(\underline{r}_1) A_1(\underline{r}_2) \ldots h(\underline{r}_1) P' A_1(\underline{r}_1) A_1(\underline{r}_2) \ldots \tag{10.60}$$

where the permutations P and P$'$ are of the type explicit in (10.12). Since the one-particle A functions are orthogonal the only nonzero terms in (10.60) are those with the same A functions (same state) fore and aft of $h(\underline{r}_k)$ for all $\underline{r}_i$ except $\underline{r}_k$. In the case of $\underline{r}_k$ the presence of $h(\underline{r}_k)$ prevents the requirement of orthogonality for the A functions of $\underline{r}_k$. It is easy to see that every term in (10.60) boils down to

$$V_{pq} = \int \psi_p^*(\underline{r})\, h_1(\underline{r}) \psi_q(\underline{r})\, d\underline{r} \tag{10.61}$$

where p and q are some integers, and $\underline{r}$ here is $\underline{r}_k$. What is hard to see is how to identify the integers p and q and how to count up the number of these terms.

Since we follow the rate of change of n_i, it is clear that one of the states p and q must be i, and we refer to the other as j, for neatness. Evidently we have to consider one-particle transitions $i \to j$ and $j \to i$ for each term of (10.60). In the case of $i \to j$ we can describe the transition as a transition for the whole system from $\underline{n}_o$ to $\underline{n}$:

$$\underline{n}_o = (n_{o1}, n_{o2}, \ldots, n_{oi}, \ldots, n_{oj}, \ldots) \to \underline{n}$$

$$\underline{n} = (n_{o1}, n_{o2}, \ldots, n_{oi} - 1, \ldots, n_{oj} + 1, \ldots) \qquad (10.62)$$

We can abbreviate this notation to

$$(n_{oi}, n_{oj}) \to (n_{oi} - 1,\ n_{oj} + 1) \qquad (10.63)$$

and we also have

$$(n_{oi}, n_{oj}) \to (n_{oi} + 1,\ n_{oj} - 1) \qquad (10.64)$$

In fact, for given $\underline{n}_o$, and $\underline{n}$ corresponding to j in (10.63) or (10.64), we can express $\bar{H}(\underline{n}_o, \underline{n})$ in (10.58) as

$$\langle n_{oi}, n_{oj}\ |V|\ n_{oi} + 1,\ n_{ij} - 1\rangle \qquad (10.65)$$

or

$$\langle n_{oi}, n_{oj}\ |V|\ n_{oi} - 1,\ n_{oj} + 1\rangle \qquad (10.66)$$

This technique for simplifying the whole-system transition to the transitions of one or a few particles is generalized by the "diagram techniques" of quantum theory. What is left after that is the harder problem of counting up the number of terms of each kind.

Let us count the terms of (10.65). Looking at (10.60) we notice that one of the sums over permutations can be omitted for the other sum will certainly catch all nonzero terms. We omit P'. Secondly we notice that the number of particles initially in state i that can have a contributing term from (10.59) is n_{oj}. Finally we notice that the number of terms that can be formed for each contributing $h(\underline{r}_k)$ is the number of permutations P, which is $W(\underline{n})$. Thus we have (10.58) as

$$\langle n_{oi}, n_{oj} | V | n_{oi} + 1, n_{oj} - 1 \rangle = n_{oj} W(\underline{n}_o) \frac{1}{\sqrt{W(\underline{n}_o) W(\underline{n})}}$$

$$= \sqrt{(n_{oi} + 1) n_{oj}}\, V_{ji} \qquad (10.67)$$

where V_{ji} is defined by (10.61). By similar reasoning we find

$$\langle n_{oi}, n_{oj} | V | n_{oi} - 1, n_{oj} + 1 \rangle = \sqrt{n_{oi}(n_{oj} + 1)}\, V_{ij} \qquad (10.68)$$

Thus we have (10.56) as

$$\frac{dn_i}{dt} = \sum_j T_{ij} \left[(n_{oi} + 1) n_{oj} - n_{oi}(n_{oj} + 1) \right]$$

$$= \sum_j T_{ij} n_{oj} - \sum_j T_{ij} n_{oi} \qquad (10.69)$$

where

$$T_{ij} = \frac{2}{\hbar^2} |V_{ij}|^2 \frac{\sin \omega_{ij} t}{\omega_{ij}} \qquad (10.70)$$

The form of (10.69) is especially interesting in the present context because it is the form of a stochastic equation and yet it has been derived from the mechanical Schrödinger equation without the intervention of any statistical considerations. It is true that we have followed, for a short time, the evolution of the system from an initial distribution $\underline{n}_o$, but nowhere have we needed to recognize $\underline{n}_o$ as an equilibrium distribution or a most probable distribution. It is a "stationary state" distribution and would not change in time without the influence of the perturbing external force, but the stationary states of quantum mechanics are not the same as the equilibrium states of statistical mechanics. The latter are determined by macroscopic fluid properties.

Our conclusion here is that the response of a system initially in a stationary state to a perturbing external influence is governed by an equation (10.69) which has the stochastic form even though it is explicitly reversible in time (the T_{ij} change sign under time reversal). The non-stochastic feature of the equation is that the coefficients T_{ij} are calculated from mechanics through (10.70) and the arguments leading to (10.51). There is also no commitment concerning the initial distribution. The equilibrium solution of (10.69)

$$0 = \sum_j T_{ij} n_{oj} - \sum_j T_{ij} n_{oi} \tag{10.71}$$

with the symmetry $T_{ij} = T_{ji}$ is the uniform distribution

$$n_{oi} = n_{oj}, \quad \text{all } i, j \tag{10.72}$$

Thus the uniform (over states) distribution is the only initial distribution that would not change under the perturbation. This is appropriate on mechanical grounds because of the symmetry of the transition probabilities T_{ij}. It is also the most probable distribution (9.24) in the absence of an energy constraint or any other constraint (except the fixed value for n).

The resemblance of (10.69) to a stochastic equation may be simply a puzzling coincidence or it may represent something deeper than that. In any case it has been found here only for the case of a system without interparticle interactions responding to a fixed external force. It will be much more interesting for us to consider a system with interactions. We will omit the external forces.

10.3 The Two-Particle Approximation

Much of the formalism outlined above was so general that it applies equally well to the two-particle approximation or any other approximation. For example, the master equation (10.52) and the transition to the occupation number representation in Eqs.(10.53) - (10.57) involve no restrictions to a particular number of interacting particles. The restriction came in (10.59), where interactions among particles were ignored. Similarly, the earliest equations of this chapter involved no restrictions.

We could start here by trying to force the general equations above to reduce to equations for two interacting particles, ignoring external forces. Perhaps we could take the wavefunction for two particles and extract the probability density for one particle:

$$|\psi^{(1)}(\underline{r}_1,t)|^2 = \int d\underline{r}_2 |\psi^{(2)}(\underline{r}_1,\underline{r}_2,t)|^2 \qquad (10.73)$$

This is similar to the way the one-particle distribution function $f^{(1)}$ is extracted from $f^{(2)}$ in classical mechanics. Unfortunately, there is no precedent or rationale for that. We cannot expect to come up easily with a BBGKY theory for reducing many-particle equations to fewer-particle equations. This may be due partly to the symmetry requirements of quantum theory, which are more explicit than those of classical theory, and partly due to the fact, reluctantly accepted, that thermal statistics is not contained in quantum mechanics any more than in classical mechanics. Quantum mechanics has its own probabilities, perhaps, but those are not the probabilities needed to reach the macroscopic level according to the method of statistical mechanics. The best clue for that distinction is that the method of statistical mechanics is always based on finding the most probable distribution of something, the equilibrium distribution, and that concept is missing in quantum mechanics. The dice of quantum mechanics are not the same as the dice of thermodynamics.

So let us start by simply pursuing the quantum description of a system of interacting particles and try to restrict ourselves to interactions involving only two particles at a time. We consider a system of N identical particles with an interaction potential energy

$$H_1 = \sum_{r>s} V_{rs} \qquad (10.74)$$

The zero-order wavefunctions are the boson functions used above and, in fact, we have the formalism developed above all the way to and including Eq.(10.58). We replace (10.59) by (10.74). Thus we have (10.58) as

$$H(\underline{n}_o, \underline{n}) = \sum_{r>s} \sum_{P} \sum_{P'} \frac{1}{\sqrt{W(\underline{n}_o)W(\underline{n})}} \cdot \int dq\, PA_1(\underline{r}_1)A_1(\underline{r}_2) \ldots V_{rs} P'A_1(\underline{r}_1)A_1(\underline{r}_2) \ldots \tag{10.75}$$

The orthogonality of the A functions weeds out most of the terms. In the surviving terms, for given r and s, only particles r and s can be in different states fore and aft of the factor V_{rs}. So we can eliminate the sum over P', as we did above, and restrict ourselves to cases where $\underline{n}$ differs from $\underline{n}_o$ only in the populations n_j of four states. If we think of the interaction between particles in states i and j as

$$i + j \rightleftarrows k + \ell \tag{10.76}$$

then we may say the distribution $\underline{n}_o$ has changed to

$$\underline{n} = (n_{o1}, n_{o2}, \ldots n_{oi} - 1, \ldots, n_{oj} - 1, \ldots, n_{ok} + 1, \ldots, n_{ok} + 1, \ldots) \tag{10.77}$$

Let us now drop the o subscripts and use the abbreviation

$$(n_i, n_j, n_k, n_\ell) \rightarrow (n_i - 1, n_j - 1, n_k + 1, n_\ell + 1) \tag{10.78}$$

There is also the inverse

$$(n_i, n_j, n_k, n_\ell) \rightarrow (n_i + 1, n_j + 1, n_k - 1, n_\ell - 1) \tag{10.79}$$

The integrals in (10.75) boil down to integrals of the form

$$V_{ijk\ell} = \int \psi_i^*(\underline{r}_1)\psi_j^*(\underline{r}_2)V(\underline{r}_1, \underline{r}_2)\psi_k(\underline{r}_1)\psi_\ell(\underline{r}_2) \tag{10.80}$$

or complex conjugates of (10.80), and the only problem remaining is to count up the number of such integrals in (10.75). The procedure is similar to the one described above in the derivation of (10.69) (Mirarchi, 1967). When we put the results of the counting into (10.53) - (10.56), we find the analog of (10.69) for the present case as

$$\frac{dn_i}{dt} = \sum_{j,k,\ell} T_{ijk\ell}[n_k n_\ell(1+n_i+n_j) - n_i n_j(1+n_k+n_\ell)] \tag{10.81}$$

where

$$T_{ijk\ell} = \frac{8}{\hbar^2}|V_{ijk\ell}|^2 \frac{\sin \omega_{ijk\ell}t}{\omega_{ijk\ell}} \tag{10.82}$$

$$\omega_{ijk\ell} = \frac{E_i + E_j - E_k - E_\ell}{\hbar} \tag{10.83}$$

The form of the result (10.81) of these calculations is especially interesting for many reasons. First of all, it does not agree with the stochastic form of a rate equation for (10.76). The nearest stochastic form is

$$\frac{dn_i}{dt} = \sum_{jk\ell} (T_{ijk\ell}n_k n_\ell - T_{k\ell ij}n_i n_j) \tag{10.84}$$

where the coefficients T_{ijkl} are parameters to be found empirically. Even if we have the symmetry

$$T_{ijk\ell} = T_{k\ell ij} \tag{10.85}$$

which can come only from mechanics, as it does in the case we consider, we still have a form

$$\frac{\partial n_i}{\partial t} = \sum_{jk\ell} T_{ijk\ell}(n_k n_\ell - n_i n_j) \tag{10.86}$$

which is different from (10.81). It is similar to the forms of the stochastic rate equations of chemical kinetics (8.61) and the Boltzmann equation, (9.225) with (9.229).

Secondly, the equilibrium solution of (10.81) should offer a new clue about the origin of irreversibility. It will not be the same as the solution of the simpler equation, (10.86), and the comparison of those solutions with each other and with the most-probable distributions of statistical mechanics could help us understand the relation between particle mechanics and fluid equilibrium.

First we notice that the equilibrium solution of (10.86) must come from

$$n_k n_\ell = n_i n_j \tag{10.87}$$

where the states identified by the indices are related by the collision

$$i + j \rightleftarrows k + \ell \tag{10.88}$$

This is equivalent to the problem solved by the H-theorem of classical statistical mechanics (9.263) expressed as

$$f_1' f_2' = f_1 f_2 \tag{10.89}$$

for the collision

$$\underline{v}_1' + \underline{v}_2' \rightleftarrows \underline{v}_1 + \underline{v}_2 \tag{10.90}$$

(Here we have evaluated (9.263) as zero, which is the equilibrium condition.) The solution of (10.89) by various classical arguments is the most probable distribution of statistical mechanics,

$$f = Ae^{-E/kT} \tag{10.91}$$

where E is the kinetic energy and the "normalization coefficient" A depends on local values of the density n and temperature T. The detailed form of (10.91) is familiar to us in a variety of contexts: (3.23), (9.67), (9.139), and (9.177). By analogy we see the discrete form of the solution (10.91) as

$$n_i = Ae^{-E_i/kT} \tag{10.92}$$

This is, of course, the well-known formula (9.58). To see that (10.92) satisfies (10.87) we only need to notice the equality:

$$e^{-\frac{1}{kT}(E_i+E_j)} = e^{-\frac{1}{kT}(E_k+E_\ell)} \qquad (10.93)$$

which is due to the equality of the energies before and after the collision (10.88).

With that digression to (10.87) in mind, let us tackle the mechanical equation (10.81) which comes from the Schrödinger equation. The mechanical equilibrium condition is described by

$$n_k n_\ell (1 + n_i + n_j) = n_i n_j (1 + n_k + n_\ell) \qquad (10.94)$$

We can carry the algebraic solution to the extent of discovering a clue, but not to completion (finding a unique solution). The clue is to replace n_i by x_i

$$n_i = \frac{1}{x_i - 1} \qquad (10.95)$$

Then we have

$$n_i + n_j + 1 = \frac{x_i x_j - 1}{(x_i - 1)(x_j - 1)} \qquad (10.96)$$

Thus (10.94) becomes

$$x_i x_j = x_k x_\ell \qquad (10.97)$$

which reminds us of (10.87). In fact, if we use

$$x_i = ae^{E_i/kT}$$

or

$$n_i = \frac{1}{ae^{E_i/kT} - 1} \qquad (10.98)$$

then we do have a solution of (10.94). Here the coefficient a is left to be determined by normalization.

Actually (10.98) is precisely what is found by the method of statistical mechanics for bosons (Tolman, 1938, page 373), where the parameter a is identified as a function of temperature and density. For low enough temperature or high enough density, a becomes so large that (10.98) becomes

$$n_i = \frac{1}{a} e^{-E_i/kT} \tag{10.99}$$

Furthermore, the values prescribed by this formula are very small compared to unity. Thus (10.81) reduces to (10.86).

It is well known that a gas of bosons at familiar temperatures has the same equilibrium distribution as a gas of boltzons. What we have found is that the rate equation for the change of the boson distribution is the classical Boltzmann equation. That presents a puzzle because the Boltzmann equation depends on the factoring of the two-particle distribution function due to the molecular chaos assumption while the quantum rate equation requires no such hypothesis. The two-particle distribution was not even mentioned in the derivation above. Is it possible that the molecular chaos assumption needed for the classical derivation is already built into the Schrödinger equation? Can quantum mechanics go all the way to the dynamics of irreversible processes at the fluid level without the introduction of statistics at any point?

It is here that we find our last clue for the origin of exclusion irreversibility. We have noticed that classical and quantum approaches to the derivations of stochastic equations are very different. But there is one common denominator, namely, equilibrium. The role of a past equilibrium in the classical derivation has been indicated in a variety of ways in earlier chapters. The role of equilibrium in reducing the mechanical equation (10.81) to the stochastic equation (10.86) was passed over too quickly above. We noticed that the n_i on the right side of (10.81) are small compared to unity at high temperatures, and that reduced (10.81) to (10.86). It should be emphasized that the argument there is based on equilibrium statistical mechanics. We have assumed that the distribution n_i was initially either the equilibrium distribution or the most probable ("local equilibrium") distribution and has evolved from that past condition for a short time. If the rate equation fails for the time-intervals of interest, then we will have to include higher-order terms.

We cannot argue that (10.81) reduces to (10.86) even at high temperatures for an arbitrary distribution. The argument can be made near recent equilibrium.

Equation (10.86) has been referred to above as the Boltzmann equation. To actually transform that equation into the classical Boltzmann equation requires a derivation in more detail than we want to consider. The claim that (10.86) is the Boltzmann equation can be supported for rigid spheres by a quick argument.

Let us introduce the delta function $\delta(r)$ defined by

$$\int dr\ \delta(r - 0)\, f(r) = f(\sigma) \qquad (10.100)$$

Introduce a potential energy defined by

$$V(r) = \sigma\epsilon\, \delta(r - 0) \qquad (10.101)$$

where σ is the sphere diameter and ϵ is an energy, a kind of barrier height for colliding particles. The time to cross the barrier at the relative speed v_{21} is (σ/v_{21}). If we assume that this time and the barrier height are related by the uncertainty principle

$$\Delta E\, \Delta t = \epsilon(\sigma/v_{21}) = h \qquad (10.102)$$

then we find

$$\epsilon = \frac{hv_{21}}{\sigma} \qquad (10.103)$$

Putting this formula into (10.101) and using that in (10.80) with free-particle wavefunctions, we can find the formula for the transition probabilities (10.82). Applying (10.49) and noticing the restriction on indices implied by (10.76), we obtain the Boltzmann equation. It is necessary, of course, to replace the state designations i and j by velocities $\underline{v}_1$ and $\underline{v}_2$ and to replace the sum over states by an integral.

10.4 The Chemical Approximation

Consider a collection of n atoms B and m diatomic molecules B_2 undergoing the chemical reaction

$$B_2 \rightarrow 2B \tag{10.104}$$

To be general we have to allow for the reverse reaction, and, in fact, many reactions, such as

$$B + B_2 \rightarrow 3B \tag{10.105}$$

Suppose the distribution of atoms over states $A_1, A_2, \ldots$ at any one time is indicated by $\underline{n} = (n_1, n_2, \ldots)$ and the distribution of molecules over molecular states $M_1, M_2, \ldots$ is indicated by $\underline{m} = (m_1, m_2, \ldots)$. We want to derive a rate equation for (10.104),

$$\frac{dm}{dt} = f(m,n) \tag{10.106}$$

from the Schrödinger equation.

The derivation can proceed as in Section 10.1 if we enlarge the list of populations n_i to include the m_i. For example, (10.10) becomes

$$W(\underline{n},\underline{m}) = \frac{(n+2m)!}{n_1!\,n_2!\,\ldots\,m_1!\,m_2!\,\ldots} \tag{10.107}$$

and (10.11) must be extended to include the B_i factors. The permutations of (10.12) also act on the atoms within the molecular pairs. It is easy to generalize all of the subsequent formulas until we come to (10.22). The problem which arises there is the "heart of the matter" in this section. Without this problem we would find nothing new here.

The problem we refer to is the problem of how to separate the Hamiltonian into a zero-order part and a perturbing part. We do not want to include all of the atomic interactions in H_1 as we did in Section 10.3. Here we want to include only the interactions among atoms and molecules, not the interactions within molecules. The interactions within molecules are represented by the chemical composition $(\underline{n},\underline{m})$ and they are present in the zero-order Hamiltonian. We can use the expansion (10.23), but we must find a way to insert the paramater λ in the formula for the Hamiltonian H for the whole system. Unfortunately there is no way to do that.

Chemical composition is a "zero-order thing," and it cannot be acknowledged in the general formula for H.

The way to introduce a "chemical perturbation theory" so that the perturbing forces are only those that cause chemical change is to delay the decision about where to insert the perturbation parameter λ in H until we have H operating explicitly on a particular term in $\psi(\underline{n},\underline{m})$ such as the term with

$$\psi = A_1(1)A_1(2)\ldots A_1(n_1)A_2(n_1+1)\ldots M_1(n+1,n+2)\ldots$$
$$\ldots M_1(n+3,n+4)\ldots M_1(n+2m_1-1,\, n+2m_1)\ldots \tag{10.108}$$

analogous to (10.11). Then we can look at the array

$$H = \sum_i K_i + \sum_{i>j} V_{ij} \tag{10.109}$$

where the K_i are kinetic energy operators, and pick out the potential energy terms V_{ij} that must be multiplied by λ. Those are the V_{ij} where particles i and j are not paired into a molecule in the formula (10.108). The same decision can be made separately for each term in the sum over permutations which generalizes (10.12).

The appearance of arbitrariness in the procedure outlined above disappears when we finally integrate as in (10.29) to obtain equations for the rates of change of the $C(\underline{n},\underline{m})$. The integrals H_{nm} are more complicated but it is clear as to which terms are perturbations and which are not. The details of the subsequent derivation (Mirarchi, 1967) are similar to the details in the derivation of the previous section, though requiring more cleverness in counting procedures.

The result obtained by Mirarchi can be described by presenting the analog of (10.81) for the predominant contribution to

$$B_2 \rightleftarrows 2B \tag{10.110}$$

due to

$$B + B_2 \rightleftarrows 3B \tag{10.111}$$

That analog is

$$\frac{dn_i}{dt} = \sum_{jk\ell p}' T_{ijk\ell p} \begin{bmatrix} (n_i - 1)n_\ell \cdot n_j n_k (n_\ell + 1)(m_p + 1) \\ -n_i^2 n_j n_k (n_\ell + 1)(m_p + 1) + \\ + (n_i + 1)^2 (n_j + 1)(n_k + 1) n_\ell m_p - \\ - n_i (n_i + 1)(n_j + 1)(n_k + 1) n_\ell m_p \end{bmatrix}$$

(10.112)

where the n_s are atomic populations, the m_s are molecular populations, and the transition probabilities are

$$T_{ijk\ell p} = \frac{24 \sin \omega t}{\hbar^2 \omega} \left| \langle i,j,k|V|\ell,p\rangle \right|^2 \tag{10.113}$$

What is of interest here is the form of the terms in the brackets on the right side (10.112). If we cancel terms there we find the bracketed expression as

$$n_\ell m_p (n_i + 1)(n_j + 1)(n_k + 1) - n_i n_j n_k (n_\ell + 1)(m_p + 1) \tag{10.114}$$

What is to be expected for (10.111) from the stochastic reasoning of the "mass action law," as illustrated in (8.57), is

$$n_\ell m_p - n_i n_j n_k \tag{10.115}$$

In fact, this replacement for the bracketed factor on the right of (10.112) offers a more detailed expression of the mass-action law than does the familiar formula

$$\frac{d}{dt}[B] = k_1 [B][B_2] - k_2 [B]^3 \tag{10.116}$$

In (10.112) we notice each separate state of B and B_2 and sum over them. If we can express (10.112) as

$$\frac{dn_i}{dt} = \sum_{jk\ell p} T_{ijk\ell p} (n_\ell m_p - n_i n_j n_k) \tag{10.117}$$

then we can say we have derived a fundamental form of the kinetic law (10.116) and can use it to calculate the rate constants k_1 in (10.116).

Our problem is that we have (10.114) from mechanics and we want (10.115) to reach chemical kinetics. Let us recall that at or near equilibrium, the boson distribution n_1 at familiar temperatures is such that n_1 is very small compared to unity. Thus (10.114) does reduce to (10.115), and we have our desired result.

Once again we find, at the end of the line, the desired stochastic formula known to prevail in many situations of fluid dynamics. And the requirement that had to be added to mechanics to get that desired formula was a restriction to equilibrium in some sense. We could argue further here, as we have earlier, that the equilibrium referred to is a recent past equilibrium.

The applications available to us when we have acknowledged our debt to recent equilibrium can be illustrated in terms of (10.117). Suppose we acknowledge (10.117) and the thermal equilibrium within the atomic and molecular species, inherited from a past equilibrium. So we have arbitrary numbers n of atoms and m of molecules (not at chemical equilibrium) but the atoms have a most probable distribution over atomic states at temperature T '9.58)

$$n_i = n\left(e^{-\frac{1}{kT} E_i^{(a)}} / q_a\right) \tag{10.118}$$

and the molecules have a most probable distribution

$$m_p = m\left(e^{-\frac{1}{kT} E_p^{(m)}} / q_m\right) \tag{10.119}$$

Then we have the chemical rate equation from (10.117) as

$$\frac{dn}{dt} = k_1 nm - k_2 n^3 \tag{10.120}$$

where the "rate constants" k_1 and k_2 are given by

$$k_1 = q_a^{-1} q_m^{-1} \sum_{ijk\ell p} T_{ijk\ell p}\, e^{\frac{1}{kT} E_i^{(a)}}\, e^{-\frac{1}{kT}(E_\ell^{(a)} + E_p^{(m)})} \tag{10.121}$$

$$k_2 = q_a^{-3} \sum_{ijk\ell p} T_{ijk\ell p}\, e^{\frac{1}{kT} E_i^{(a)}}\, e^{-\frac{1}{kT}(E_i^{(a)} + E_j^{(a)} + E_k^{(a)})} \tag{10.122}$$

It is possible that putting the molecular model (formula for V_{ij}) into these formulas will give good values for the rate constants. They should be accurate for predictions based on initial thermal equilibrium in the absence of chemical equilibrium. We could say that equilibrated samples of B and B_2 were mixed initially and then allowed to react.

In conclusion we find that the assumption of recent equilibrium is the ingredient to be added to particle mechanics, classical or quantum, to lead the mechanics through statistics to a prediction of fluid dynamics. It is never possible to go all the way with mechanics alone, although we can come close, especially in quantum mechanics. There is always a need for the acknowledgement of equilibrium in the recent history of the system. This need may not be so strong at low temperatures, where we have nonstochastic quantum mechanical equations involving only one-particle distributions, and/or one-molecule distributions. But those equations (nonlinear) are hard to solve and we have no reason here to explore their range of applications. We have found what we wanted, namely a general dependence of fluid predictions on a recent equilibrium at familiar temperatures at gas densities. It is desirable to push the theory beyond those restrictions to low density and high temperature, but no attempt is made here to pursue the subject in those directions. We are content to notice that the mass-action law, like other stochastic laws, can be explained in terms of recent equilibrium, and the explanation is good enough to provide formulas for rate coefficients.

CHAPTER 11

ON ALTERNATIVE APPROACHES

Having completed our discussion about the nature and origin of irreversibility, it is time to step back and assess the result. The mechanism we offer has the virtue of simplicity, but on the other side, it may seem to be too simple to account for the pervasiveness of irreversible phenomena. For example, we have noticed that many theorists have sought the source of irreversibility in the domain of the very large or of the very small. They have tried to find macroscopic or microscopic mechanisms which provide a basis for irreversibility. Translated into the context of contemporary physical theory, this means that cosmology or quantum mechanics is expected to yield a mechanism for the prevalent irreversibility of everyday life.

The cosmological origin has been advocated most eloquently by Gal-Or (1974, 1975, 1982) and also by others, such as Gold (1958, 1967, 1974) and P.C. Davies (1974, 1977). The kind of thermodynamic thinking that characterizes their approach is the following. Empty space is very cold because of the expansion of the universe and this coldness provides a radiation sink which prevents sections of the universe from reaching thermal equilibrium. Thus the expansion of the universe is the ultimate origin of all irreversibility.

Others disagree and claim that the expansion of the universe is irrelevant since it is, in principle, reversible, and if a contraction phase were to occur it would not affect the fusion processes of stars. Maybe the primary source of disequilibrium is the fusion processes of stars. And other cosmological thinkers reject both of these mechanisms (expansion and fusion) as the origin of irreversibility and focus instead on self-gravitating systems which are said to possess "an infinite reservoir of negative entropy." According to Davies (1974, p.108), "the origin of all thermodynamic irreversibility in the real universe depends ultimately on gravitation."

While we appreciate the scale and grandeur of the cosmological approach, we are doubtful that the source for <u>all</u> irreversibility is to be found here. First, in spite of the wide acceptance of the "big bang" model of the universe, it must be remembered that this <u>is</u> a theory and not so highly corroborated that alternatives are unimaginable. There is evidence for this model but the evidence can be interpreted in other ways. But even if this cosmological model had the assuredness of other areas of science, we would look for some demonstrations of strong coupling between cosmological processes and intermediate phenomena. Just how does universal expansion, stellar fusion or gravitational collapse affect everyday processes of diffusion and heat transfer? In what way do these latter phenomena inherit irreversibility from the cosmic level?

In absence of explicit mechanisms which show this inheritance, the irreversibility rooted in cosmological processes cannot be assumed to be the progenitor of all other kinds of irreversibility. It seems to us that the present state of knowledge does not warrant the move from the cosmological to all other levels. We have many theories and not very many demonstrated mechanisms. Given the many theories and the lack of a set of concepts which would establish continuity from intermediate to cosmological levels, we are skeptical of claims that would trace all irreversibility to cosmic sources.

From another direction we have the suggestion of a new approach to particle and fluid physics. Rather than look outwards for more distant data, let us, we are told, look inwards at our old schemes of particle dynamics. Perhaps Newtonian mechanics and its enlargements, which dominate current physics, are too static in their stone-cold determinism to fully explain the order that arises out of disorder in large collections of particles, especially in biological systems. They are missing an essential ingredient which, according to Prigogine (1980), is just now being uncovered. The new ingredient will not contradict old theories; it will supplement them.

There are new discoveries about how order comes out of chaos and the excitement about them has reached the popular press (Gleick, 1984). Gleick refers to the unpredictability of well-defined weather patterns, the mysteries in the physiology of the heart rhythm, the blunders of economics, the interesting patterns of swirling cigarette smoke, and a number of other familiar phenomena. Quoting highly respected mathematicians he points to their surprise when models seem to suddenly reveal potentialities for behavior that was not anticipated. Certain values of initial data sometimes lead to surprising variations at late stages in the evolution. Periodic solutions sometimes occur where least expected.

In the spirit of this new view we could argue that it is not enough to see life-order as emerging from a computer-like program that causes organisms to reach out for frequent inputs that interrupt the devasting pull toward equilibrium. There must be some intrinsic tendency of disordered matter to organize itself. It is not enough to see accidental mutations and severe selections as the total mechanism of biological evolution anymore than it is enough to see probability as the total basis for fluid organization. (We are not referring here to recent theories about biological evolution which find sociological aids to selection or isolations of gene pools but to more drastic tendencies among the genes to improve themselves.) There must be, in the spirit of this new view, an "organizing force" which has not yet been discovered but which is about to emerge, maybe even in physics.

In the case of biological systems, we could argue that aging is a loss of "software" which leaves the "hardware" more and more at the mercy of dissipative processes, and that much of the view is traditional. But the process of regeneration in such systems is seen as proof of a natural tendency toward order. Evidently, the gene pool continually gains more in order and sophistication than it loses to the forces of decay. This cannot be due entirely to the slow process of accidental mutations and selection.

For our purposes it is desirable to pass by the more dramatic speculations about biological systems and consider the "new trend" as it may be developing in physics. We refer to the recent attention given to the emergence of complex and irreversible patterns of flow and other kinds of change at the macroscopic level in fluids that are not expected to yield such orderly patterns. The molecular motions in the fluids are highly chaotic, and it is surprising that so much order can arise out of so much chaos. Prigogine (1980) refers to the macroscopic patterns as "coherent" or "dissipative" structures and finds their origin in certain kinds of instabilities. He offers many very interesting examples. The exciting feature of his assessment is the suggestion that the explanation of these seemingly spontaneous and maybe even evolutionary phenomena requires a revolution in physics that will bring the subject alive and will introduce revitalizing modifications at the most fundamental levels. The seeds of irreversibility will become recognizable at the level of molecular dynamics, or at some new level between molecular dynamics and fluid dynamics. The origin of irreversibility cannot be identified until we have first produced some new physics.

This is not to say that the study of all ordered patterns of convective motion in fluids is new. It is well known, for example, that a liquid heated from below in gravity can develop complex patterns of convection, e.g., the famous Benard "cells." In fact, any kind of convection represents an emergence of orderly fluid motion out of the chaotic molecular motions. This is the mystery of the fluid concept described in Section 8.1. What is new is a sudden jump in our awareness of the mystery and in the vast variety and complexity of its manifestations. We may very well be at the edge of an exciting decade of discoveries about the potentialities of fluids for sustaining "coherent" structures, but we do not know where to draw the line between the mysteries of the fluid concept, or for that matter, the fluid condition, which attends "classical" phenomena, and the new mysteries involved in more spectacular phenomena. It is possible, of course, that the science of the fluid emergence, when developed, will not draw any such line.

And the other thing that is new is Prigogine's suggestion that the new physics will involve modifications at a fundamental level to allow for extensions of particle dynamics to new non-Hamiltonian (non-Newtonian) forms that will admit temporally irreversible components to become the roots of temporal irreversibility in fluid phenomena. The idea is that such new and irreversible components could be the underlying cause for the new and spectacular phenomena observed at the macroscopic level.

Here we have to disagree. In the first place, we do not believe in "temporal irreversibility" under any circumstances, as explained in Chapter 5, and we do not go for "entropy operators" or any such excess formalism, or "transformation theories" that bring in new physics, or conclusions that depend on "mixed" states as opposed to "pure" states, or conclusions that require "continuous spectra" of operators to "discrete spectra," etc. These parts of Prigogine's arguments seem to us to be somewhat vague and unnecessary, maybe even irrelevant. But the whole of his argument is most stimulating and helpful. We are led from that to make our own arguments as follows.

First of all, let us consider the simplest emergence of pure and orderly fluid convection out of molecular disorder. We have an ideal gas which has come to equilibrium uniformity with the familiar velocity distribution

$$f = n\left(\frac{m}{2\pi kT}\right)^{3/2} e^{-\frac{m}{2kT}v^2} \tag{11.1}$$

If this gas just reached equilibrium in recent times, it would just sit there without any macroscopic activity for very long times into the future. But let us suddenly impose, at the initial time, an external force, and in this case let the force be the force of gravity. In other words, the acceleration on every particle is

$$\underline{a} = (0,0,-g) \tag{11.2}$$

where g is the gravitational constant, 9.8 m $\sec^{-2}$. The trajectory of a molecule initially at $\underline{r}$ with velocity $\underline{v}$ would be

$$\underline{v} = \underline{v} + \underline{a}\,t$$
$$\underline{r} = \underline{r} + \underline{v}t + \frac{1}{2}\underline{a}t^2 \tag{11.3}$$

and so we have the solution of the Liouville equation analogous to (9.97) as

$$f(\underline{r},\underline{v},t) = f\left(\underline{r} - \underline{v}t - \frac{1}{2}\underline{a}t^2,\ \underline{v} - \underline{a}t, 0\right) \tag{11.4}$$

Putting this into (9.174) and (9.175) we find the fluid properties at time t for a rarefied gas. It turns out that the property densities remain unchanged and no conductive fluxes have developed but the gas is moving convectively in the positive z-direction with velocity

$$\underline{v}_o = (0,0,gt) \tag{11.5}$$

The purely convective heat flux is

$$\underline{q} = \left(\frac{3}{2}nkT + \frac{1}{2}mv_o^2\right)\underline{v}_o \tag{11.6}$$

If we terminate the flux growth at the mean collision time, we have in these formulas the molecular interpretation of buoyancy in collision-dominated gases.

The molecular description of buoyancy given above can be easily extended by introducing boundary conditions that lead away from uniformity to the equilibrium barometric equation

$$n(z) = n(0)e^{-\frac{mgz}{kT}} \qquad (11.7)$$

We will not pursue those details. Our purpose here is merely to offer a simple example of how orderly convection can come out of the molecular chaos. To understand the cause here we start with the initial condition where the molecules are moving randomly with the Maxwellian distribution of speeds. The imposition of gravity has a very small effect on each molecular trajectory, accelerating it downward for a very short time (a mean collision time). That causes a very slight redistribution of the molecular velocities. Faster molecules are decelerated for less time and tend to "move upward" relative to slower molecules leading to an upward convective motion for the fluid as a fluid. The molecular collection retains its randomness, for the most part, and what we see at the fluid level is an orderly change coming out of the order (not disorder) of the original equilibrium randomness. We are reminded here that "order" in the sense of a definite collection of small effects is the brother (not the enemy) of "chaos," which cancels out everything except for the small effects.

The argument above can be applied to every fluid pattern of change, including (especially) the equilibrium pattern of no change, which is coherent and reproducible. It is not the fluid behavior that must be seen from a new point of view but rather our prejudice that must be revised. We tend to see conduction as a reasonable consequence of molecular randomness while convection is more surprising. And the convection, if it involves "coherent" irregularities, is all the more surprising. What could be more surprising, when we think about the molecular chaos, than the swirling of cigarette smoke or the flickering of a candle flame. But those do not imply that the irregularities of convection are rooted in stochastic modifications of the underlying dynamics. They are rooted in the interactions with surroundings. Cigarette smoke into a vacuum would instantaneously fill the container without any swirls. A candle burning in a quiet atmosphere without gravity, as observed by astronauts, produces a glowing sphere. The molecular dynamics is the same in the various situations. It is the external conditions which are different. What is really surprising about fluid behavior is that there is such a thing at all. How can "fluidity" come out of the molecular chaos? That is certainly an emergence of order out of chaos! Any emergence beyond that one is easily explained by current physics.

Part of the mystique of the "coherent structures" is that they can occur far from equilibrium. But "far from equilibrium" is an ambiguous concept. It could mean that the fluid property values are far from the values they will have at a future equilibrium even though the distribution of molecules over molecular states is still near the most probable distribution dictated by the fluid property values. In that case, we can still use the fluid equations, including equations of flux and equations of equilibrium, to describe the fluid behavior. In this context, "far from equilibrium" is misleading. If, on the other hand, the molecular distributions are nowhere near the most probable distributions, and maybe even totally unknown, then "far from equilibrium" can be interpreted literally and the appearance of "coherent structures" in the behavior of such systems would certainly call for a new look at old physics.

Consider, for example, the case of two chemical compounds A and B mixed at an initial time after having been separately equilibrated. Looking at the mixture we could say that the system is initially at equilibrium (the past separate equilibria) or we could say the system is very far from equilibrium (the future chemical equilibrium). The rate of the chemical reaction would be highly predictable early on because of the nearness in time to a previous equilibrium. We could even use the rate equations of chemical kinetics. But the closer the system gets to the future equilibrium (the farther it goes from the past equilibrium), the less assurance we have in the validity of our predictions. This could be, and evidently is, compensated by a decreasing need for accuracy as the system approaches the new equilibrium.

The challenge presented by Prigogine (1980) is illustrated heavily by examples from chemical kinetics. He considers examples that involve hystereses, oscillations, other kinds of periodicity, and other phenomena that seem to deviate from classical notions about approaches to new equilibria. The emphasis there is, we think, very important, and it should open up a new chapter of physics. The point where we disagree is the idea that the new chapter must affect older chapters. We feel that his examples all assume the validity of the macroscopic rate equations of chemical kinetics, which, in turn, get their validity from the joining of classical physics to the assumption of recent equilibrium in the manner described above in Section 8.5. The phenomena described by Prigogine (illustrated below) are very interesting and will open up new vistas, but there is no indication from them of a need for modifications of particle dynamics and statistical mechanics.

The problems with chemical kinetics which have been identified in recent times by Prigogine and others can be described in terms of the theory developed by Poincaré (1892) and Bendixson (1901) known now as the Poincaré-Bendixson theory of autonomous differential equations (Coddington and Levinson, 1955, Chapters 13,15,16). The rate equations of chemical kinetics are often nonlinear and always autonomous. By "autonomous" we mean that the time derivative of an unknown concentration x_i is always expressed as a function f_i of concentrations that does not depend on time in any way except through the concentrations themselves

$$\frac{dx_i}{dt} = f(x_1, x_2, \ldots.) \qquad (11.8)$$

In other words, f_i does not depend directly on time t but rather is a formula involving only the x_i.

Consider for example the chemical equation

$$A + B \underset{k_2}{\overset{k_1}{\rightleftarrows}} C + D \qquad (11.9)$$

where the k_i are the "rate constants" for the forward and reverse reactions. At equilibrium we have equal rates forward and reversed,

$$k_1 [A] [B] = k_2 [C] [D] \qquad (11.10)$$

where the symbol [A] refers to the concentration of A. If we start with initial concentrations of a, b, c, and d for A, B, C, and D, and then allow for an increase x in the concentration of C, we must also allow for an increase x in the concentration of D and similar decreases in the concentrations of A and B. In other words, we have, at any time t,

$$[A] = a - x \qquad [C] = c + x$$

$$[B] = b - x \qquad [D] = d + x \tag{11.11}$$

Thus the rate equation for the rate of producing, say, C, is

$$\frac{d[C]}{dt} = k_1 [A][B] - k_2 [C][D]$$

or

$$\frac{dx}{dt} = k_1(a - x)(b - x) - k_2(c + x)(d + x) = (k_1 ab - k_2 cd)$$
$$- [k_1(a + b) + k_2(c + d)]x + (k_1 - k_2)x^2 \tag{11.12}$$

This equation is nonlinear because x^2 appears on the right side and it is autonomous because t does not appear explicitly anywhere on the right side.

The equation (11.12) is solvable and the formula appears in all texts on physical chemistry or chemical kinetics. It describes a smooth approach to equilibrium where the concentrations cease to change and obey the ratio

$$K = \frac{k_1}{k_2} = \frac{[C][D]}{[A][B]} = \frac{(c + x)(d + x)}{(a - x)(b - x)} \tag{11.13}$$

which is the "equilibrium constant." What we see here is merely (11.10) at equilibrium where the forward and reversed reactions have the same rate.

The role of equilibrium in the evolution of the system can be emphasized by stipulating that a, b, c, and d are equilibrium concentrations and x is a measure of the derivation from equilibrium. Then x_0 can be identified to provide the initial concentrations. In that case the first term on the right side of (11.12) is zero and the solution x(t) approaches zero (equilibrium).

For more complex systems we may have two or more independent but coupled chemical equations with their forward and reverse rate coefficients, as for example,

$$A + B \underset{k_2}{\overset{k_1}{\rightleftarrows}} C + D$$

$$C + E \underset{k_4}{\overset{k_3}{\rightleftarrows}} F + G \qquad (11.14)$$

Here we may introduce the final equilibrium concentrations a, b, c, d, e, f, g and the variables x_1 and x_2 as deviations back away from equilibrium for the two reactions. In other words, we could say that the concentrations at time t are

$$[A] = a - x_1 \qquad [E] = e - x_2$$

$$[B] = b - x_1 \qquad [F] = f + x_2$$

$$[C] = c + x_1 - x_2 \qquad [G] = g + x_2$$

$$[D] = d + x_1 \qquad (11.15)$$

If we express the rate equations for x_1 and x_2 in the form of (11.12) with the first term missing (equilibrium concentrations a, b, c, d, e, f, g), we find

$$\frac{dx_1}{dt} = a_{11}x_1 + a_{12}x_2 + g_1$$

$$\frac{dx_2}{dt} = a_{21}x_1 + a_{22}x_2 + g_2 \qquad (11.16)$$

where the matrix of coefficients is

$$a_{11} = -k_1(a+b) - k_2(c+d)$$

$$a_{12} = k_2 d$$

$$a_{21} = k_3 e$$

$$a_{22} = -k_3(c+e) - k_4(f+g) \tag{11.17}$$

and the nonlinear "perturbations" are

$$g_1 = k_1 x_1^2 - k_2 x_1^2 + k_2 x_1 x_2$$

$$g_2 = k_3 x_2^2 - k_3 x_1 x_2 - k_4 x_2^2 \tag{11.18}$$

In the neighborhood of equilibrium, where the x_i are small, the trajectories $x_i(t)$ are dominated by the linear part

$$\frac{d}{dt}\underline{x} = \underline{\underline{a}} \cdot \underline{x} \tag{11.19}$$

where $\underline{x}$ is (x_1, x_2) and $\underline{\underline{a}}$ is the matrix (11.17). Thus by studying the linear part alone (the matrix $\underline{\underline{a}}$) we can classify the patterns of behavior exhibited by the system (Coddington and Levinson, 1955, Chapter 15).

For example, if the matrix $\underline{\underline{a}}$ can be diagonalized and has real negative eigenvalues, the trajectories (x_1, x_2) come toward the equilibrium point. We could say the equilibrium point is an "attractor;" the equilibrium is "stable." If the eigenvalues are positive, the equilibrium point (unstable) is a "repeller." For other cases, the trajectory (x_1, x_2) moves in circles around the equilibrium point (periodic behavior) or spirals into the point or away from the point. (The ignored nonlinearities can change a circle into a spiral.) And there are other possibilities.

In the case of the reaction considered above, it can be shown that for any positive values of the rate constants and equilibrium concentrations, the matrix has two real negative eigenvalues, so the equilibrium point is a simple attractor. It is also possible to show that there is only one equilibrium point that can have positive concentrations (Coddington and Levinson, 1955, p.398). The same conclusions can be derived for any pair of coupled chemical equations like (11.14) that can have come up in the chemical kinetics of a generation ago.

Confronting us now is the use of mechanisms quite different from (11.14) which can have matrices $\underline{\underline{a}}$ analogous to the one in (11.19) but very different in certain respects. The new matrices may not have negative real eigenvalues. The solution trajectories can be repelled from the equilibrium point, or kept at bay in various ways, including periodic ways, or spiral inward toward equilibrium or outward away from equilibrium. The full variety of possibilities according to the mathematical theory of autonomous equations must now be explored. The urgency for that is all the more apparent when we notice that economists, ecologists, and others are using the kinetic models for a wider variety of phenomena. In no case, even in the cases of chemical reactions, is there a suggestion that the models do not follow the prescribed rate equations, which come from the wedding of molecular dynamics and recent equilibrium, as described above, and so there is no hint of a need for revisions in the dynamics of Newton or Schrödinger, but there is a new chapter of kinetics to be developed around the possibilities listed in the theory of autonomous rate equations.

To illustrate this need for a wider view toward the applications of the mathematical theory, we consider a model reviewed by Prigogine (1980, p.95). Suppose we have a reservoir which supplies a constant concentration of reactant A, and A reacts with a catalyst to produce even more of the catalyst,

$$A + X \xrightarrow{k_1} 2X \tag{11.20}$$

This is followed by the subsequent reactions

$$\begin{aligned} X + Y &\xrightarrow{k_2} 2Y \\ Y &\xrightarrow{k_3} E \end{aligned} \tag{11.21}$$

This example, as pointed out by Prigogine, is also used in ecological modeling where X is the density of a herbivore that uses A to reproduce and Y is a carnivore that propagates at the expense of the herbivore. But whether X is a herbivore or a chemical catalyst, we have the rate equations

$$\frac{d}{dt}[X] = k_1[X] - k_2[X][Y]$$

$$\frac{d}{dt}[Y] = k_2[X][Y] - k_3[Y] \qquad (11.22)$$

where k_1 is the true k_1 multiplied by the constant [A]. The equilibrium concentrations obtained by setting the rates equal to zero are

$$[Y]_{eq} = (k_1/k_2) = a$$

$$[X]_{eq} = (k_3/k_2) = b \qquad (11.23)$$

If we introduce displacements from equilibrium defined by

$$[X] = b + x_1$$

$$[Y] = a + x_2 \qquad (11.24)$$

then the rate equations assume the form of (11.16),

$$\frac{d}{dt}\underline{x} = \underline{\underline{a}} \cdot \underline{x} + \underline{g} \qquad (11.25)$$

where the matrix for the linear part, analogous to (11.19), is

$$\underline{\underline{a}} = \begin{pmatrix} 0 & -k_3 \\ k_1 & 0 \end{pmatrix} \qquad (11.26)$$

and the nonlinear perturbations form the vector $\underline{g}$ with components

$$g_1 = -k_2x_1x_2 \qquad g_2 = k_2x_1x_2 \qquad (11.27)$$

The eigenvalues of the matrix (11.26) are given by

$$\lambda^2 + k_1 k_3 = 0$$

$$\lambda = \pm i \sqrt{k_1 k_3} \qquad (11.28)$$

and hence the trajectories $\underline{x}(t)$ are periodic in the neighborhood of equilibrium. There is no pull toward equilibrium unless a small pull can be induced by the perturbing nonlinear part. In the present example, the vector $\underline{g} = (g_1, g_2)$ is as often outward as inward at equal distances from equilibrium, so the solution cannot be an inward spiral. The outward spiral is ruled out by the Poincaré-Bendixon theory, which predicts, for cases like this, periodic solutions. For this particular case, the periodic orbits are illustrated by Prigogine.

The results of kinetic models like the one just described may seem a bit strange to the classical kineticist, but they are certainly within the range of phenomena covered by his models, based on kinetic equations. Following the advice of Prigogine we should be prepared to expand our expectations in this area and to include effects of spatial variations (diffusion equations), and, in fact, include the whole machinery of fluid dynamics. This machinery has been pushed toward familiar phenomena without trying to explore the whole range of phenomena that are predictable within the scheme. We have been preoccupied with the building of models for familiar phenomena and have not looked beyond that to models for yet-undiscovered phenomena.

The question left for us here is whether we need to look beyond the well-worn applications of fluid dynamics to the newly emerging phenomena and go back to the molecular dynamics to infuse new elements there to render the fluid dynamics capable of handling the new phenomena. Perhaps there is a special need for new elements in the molecular dynamics which can be the seeds for irreversibility in fluid behavior.

In our opinion there is not yet any evidence pointing to a need for a revision in the schemes of Newton and Schrödinger, especially for a revision that could introduce irreversible elements for the purpose of explaining irreversibility in fluid behavior. There is certainly no need for new machinery aimed at explaining <u>temporal</u> irreversibility.

Introducing seeds of temporal irreversibility at the most fundamental levels of physics is tantamount to going against the purpose of physics, in our opinion. Even if there were such weird elements in the world (not invariant under time reversal), they would have to be held up as weird and left to be explained. There can be no starting point for physics if physical time is elevated to a level where it can "have a mind of its own." The elaboration and extension of concepts about time at higher levels, where epiphenomena have to be considered, where history and psychology come in, where philosophy seeks the final word, is another matter. But for physics, let us avoid drastic modifications of time until absolutely necessary.

APPENDIX

SOME REFLECTIONS ON TIME AND TEMPORALITY

The attentive reader will probably have many questions in mind at this juncture. We have come quite far along in our discussion about irreversibility without giving much consideration to questions about the nature of time, temporal anisotropy or temporal awareness. Our approach has been straightforward, some might even say naive. We now want to engage the reader in a different level of inquiry, one which is critical and reflective in that it aspires to make explicit some of the philosophical foundation underlying the inquiry so far. Beyond this we want to mention and briefly describe some other possible orientations towards time and temporality (Note 28).

In order to make time genuinely problematic it is first necessary to get a sense of how the purely structural concept of time emerged in Newtonian thinking. By tracing this history we seek to neutralize the sort of dogmatism which prejudges as necessary and universal that which might be contingent and historical. Our understanding of time can only be enhanced by this widening of context while any effort to see how time emerges from temporality is inseparable from cultural history.

How does a concept of time arise from the awareness of phenomena? How does a concept of time as separable from natural process originate? How does a concept of linear time evolve, and given that concept, how does a concept of homogeneous, isotropic time gain credence?

We propose no simple answers to these questions. Instead, we will use them as points of references, as twigs around which we construct a web of history. Patterns which connect contemporary views of time and temporality with their origins in other world views should then emerge.

(i) When do we begin to weave the web, around what twig do we cast the first strand? If we look to the past and look far enough then we must look at the origins of time-concepts in religion and myth. This is a vast field and it would be presumptuous to try to survey all the important developments. For the present purposes, let it simply be noted that a most compeling thesis is that "religion has stemmed from man's consciousness of Time, and that his reaction to Time has found a variety of expression, including the deification of Time." ("The Deification of Time," p.370, S.G.F. Brandon in The Study of Time, edited by J.T. Fraser et al.; see also History, Time, and Deity by Brandon.) The underlying premise is that past, present and future comprise man's existential or phenomenal temporal sense.

If the "deification of Time" thesis is accepted, then several patterns begin to emerge. From his experience of finitude, of change and decay, his sense of past, present and future man develops a concept of Time as primal. Time is the uncreated, "both Creator and the Lord of the creation which it created," says the Rivayat (Fraser, 1972, p.376).

Thus, from primordial temporality, a concept of Time as deity emerges. A cross-cultural pattern is evident, although the particular form and characteristics of the deity vary. Interesting ambivalences can be seen: in the Indo-Iranian tradition, Infinite Time conceived as ultimate reality is contrasted with time of the long Dominion, finite time which brings decay and death to all living things. With cultural diffusion, different aspects of the Time-deity are stressed, so that each concept molds and is molded by its cultural niche.

In the Judeo-Christian tradition, the deification of Time is ruled out by monotheism and an emphasis on the personal nature of God. But Time is crucially important. As Brandon notes: "... Judaism and Christianity have elaborated what are virtually philosophies of history, in that they have identified the passage of Time with the revelation and achievement of the purpose of God ..." (Fraser, 1972, p.380).

For Western theologians, the locus of the problem was redemption, the plan of God, and sacramental history. Augustine espoused an archetypal view of time in which it is represented as a completed pattern or plan. Man's finite understanding only allows him to experience this completed pattern as a succession and he must depend on seeing the sacred "prefigured" in the profane. By conceptually tying "time" to a completed plan or pattern, Augustine helped to set the state for the later secularization of nature: the view of nature as artifact and, with the rise of mechanical explanation, nature as clockwork. Nature as God's artifact is passive and inert: it has no intrinsic powers or dynamism.

(ii) There is, of course, between Augustine and Newton a large gap. The development of biological and cosmological simulacra (see J. de Solla Price, 1964, p.10, and S. Bedini, 1964, pp.24-42) and their use by scholars for illustration and argument were parts of the process of secularizing nature and history. This was a complex, multi-leveled process and one should not impose too rational a structure upon it. One has only to notice with Whitrow (Fraser, 1972, p.17), that throughout the medieval period there was conflict between the linear and cyclic concepts of time. The latter concept was favored by the scholars and scientists because of their interest in astrology and astronomy while the mercantile class associated time with money and thought of using time economically. For them time, money, and a linear progressive sense of activity form a sort of conceptual gestalt. Between this gestalt and the austere Newtonian view of "Absolute, true, and mathematical time" which "of itself, and from its own nature, flows equally without relation to anything external," no deductive line can be traced. We are content for the present purposes to notice general patterns and influences.

There seems to be a process of successive abstraction involved in arriving at the concept of homogeneous, continuous, isotropic time. If this process is seen as inseparable from cultural, social, and technological contexts, rather than "given" in some sense, then what is needed is some sense of the way cognitive gestalts are achieved. That Isaac Barrow, who preceded Newton at Cambridge, was prompted to study time by Torricelli's work in kinematics seems far removed from considerations about mechanical clocks and conceptions of God and nature. But in order for an idea to capture the imagination of an entire culture, it must be deeply rooted in the ultimate concerns of that culture.

The abstraction which taps the deepest concerns, which harmonizes and integrates them, will eventually be taken as a description of fundamental reality. In the 17th century the secular and sacred begin to diverge but this divergence was possible in part because of the convergence of social, technical, and cognitive flows. There was no sharp break with the religious world-view but rather a gradual shift in emphasis, a re-interpretation of nature. The Augustinian idea of prefiguration fades as nature as artifact, as clockwork takes hold. The mechanistic world view gets its formal embodiment and confirmation in Newton's mechanics and its associated conception of linear, homogeneous time.

In an odd way, this formalization made possible a "resacralization" of time. The absolute, true and mathematical time which flows equally without relation to anything external clearly transcends the natural world and the religious implications of this did not escape Barrow and Newton. One might speculate that Newton's rendition of the time-concept is a step down the path leading to the deification of time. Thus, Western culture in achieving science would in a sense be returning to a concept of time which monotheism and the concern for redemption and sacramental history had occluded.

For now, let us simply note that classical mechanics was fully articulated without any revision in Newton's concept of time. Indeed, classical mechanics takes the form of a deductive system and time, symbolized as "t" has no dynamical significance.

(iii) This view remained unchallenged until the development of thermodynamics and evolutionary theory. The cultural sense of linear progress which harmonized well enough with an etherealized Newtonian conception of time was undermined in the following way:

> Darwin displaced theological, teleological time and by doing so forced a cultural revision of the sense of history. Prior to Darwin, the gap between Newtonian, mathematical time and the cultural sense of teleological time was so great that no cognitive dissonance could be felt. Newtonian time is so abstract that it was possible for the culture to retain the notions of theological, teleological time, notions which reinforced the secular sense of linear progress. What Darwin did was to bring back into the world, to make it real again and put natural history at the center. If we say that Newton deified time, then Darwin secularized or naturalized time.

This naturalization finds a sympathetic resonance in the development of thermodynamics. Because of the scope of its principles and their, originally, nontechnical expression, thermodynamics influenced speculations in biology and physiology as well as social, political, psychological, historical, and religious domains. [For one of the least inspiring examples, see Henry Adams' "The Rule of Phase Applied to History" in Olson (1971).]

For our purposes, it should be noted that much of the response to the first law of thermodynamics, "the energy of the universe is constant," took the form of rhapsodies to religion and science. [See "Thermodynamics and Religion," E. Hiebert in Olson (1971).] This, however, was countered by the nontheistic, monistic, humanitarian view of people such as W. Ostwald and S. Arrhenius. Any reader who doubts the seriousness of their concern need only look briefly at Ostwald's program based on energeticism.

Ostwald writing in 1913 saw all of history as a struggle toward a unitary, monistic conception of the world. "The whole culture-evolution which has led us from the invention of the slingstone, the lever, (and) fire ... to a modern giant steamer ... signifies nothing further than an always finer and more multiform manifestation of the energetic imperative. The same may be said of our moral evolution" (Hiebert, in Olson, 1971, p.170). The energetic imperative referred to here is to waste no energy; turn it all into account.

Ostwald believed that science was naturally expanding at the expense of religion and that eventually the domain of religion would shrink to zero. His humanitarian views are evident when he says: "That we practice mutual kindness and love is no longer the demand of a Godhead standing outside of ourselves ... but it is a demand of scientific intelligence" (Hiebert, in Olson, 1971, p.170).

Thus, the first law stimulated attempts both to harmonize science and religion and to assert the primacy of science over religion.

The second law, that "the entropy of the universe tends towards a maximum," inspired even more fervent debate. This is not surprising if one interprets the law as saying that there is a unidirectional movement toward dissipation of energy, a progressive disorganization which eventuates in the "heat-death" of the universe. Such an interpretation is clearly eschatological in its substance and import and this is supposed to be the domain of theology.

We find remarks about thermodynamics and religion in the writings of Mayer, Joule, Tyndall, Kelvin, Arrhenius, and Ostwald. The latter stressed the spontaneous irreversibility of time and read it in terms of man's cultural, moral, and scientific progress. He dealt with the dystopian aspects of the second law by ignoring them. Kelvin excluded living organisms in his rendition of the law and Arrhenius thought that rare exceptions might occur which would allow for a rebuilding of worlds.

(iv) The current situation. Compared to the writings of Ostwald, Eddington, Haldane and even Boltzmann, contemporary discussions of time and entropy are quite technical and restrained in style. Why? One major reason is the influence of logical positivism on philosophy in general and philosophy of science in particular. With the discrediting of "metaphysics" and the relegation of philosophy of science to the role of handmaiden to science, the scope of discussion was narrowed while the technical and analytical depth increased. Questions about historical and cultural origins of concepts were considered outside the scope of inquiry and those philosophers interested in questions which could not easily be formulated in "scientific terms" were excluded from the discussion. The proper object of investigation was assumed to be the time of physics, that is, the concept of time used in the latest and best physical theory. Thus, temporality was acknowledged only as a subjective (perjorative sense) phenomena.

Relativity theory was the latest and most glamorous physical theory which seemed to have philosophical significance and it also was a theory which seemed in need of interpretation and analysis. Quite naturally philosophers of the logico-mathematical sort were drawn to this theory. The work of Reichenbach (1958) is representative.

There is also the strand of thought represented by what might be called the philosophers of nature. We are thinking here of people such as Bergson, James, and Whitehead who are distinct from the philosophers of science and the logico-mathematical sort because they are not scientific dogmistists. They do not grant a primacy to formal representations and have a stronger sense of science as an evolutionary development. They also have a larger philosophy of nature within which their conceptions of science and human being are embedded. For these thinkers, philosophy is natural philosophy and questions about sensory awareness, reflective awareness, perception, and temporality are very important.

As the reader might expect, the formalists (as we shall call the philosophers of science who take all their cues from scientific theory conceived as a logical organon) and the natural philosophers have conflicting views about time. The formalists argue that there is no real passage or becoming while the natural philosophers claim it is the essence of what is real. We shall not try to refine the positions or rehearse the arguments at this point (Note 29). But it is worth noting that despite their rarefied form, the contemporary positions are quite analogous to those taken in the debate about the transcendence or immanence of time which punctuated our historical sketch. When the formalists talk about the sense of passage being a consequence of our mind crawling along the world-line of its body, one can hear the echo of an archetypal view of time and history where temporal passage is merely the unfolding of an already constituted design or plan.

We feel no obligation to choose between the "formalist" and "naturalist" view about time. Our sense is that time is not simple but that for the purpose of certain kinds of analyses, one can adopt a limited concept, and work with it in hopes of revealing interesting complexities. And this is what we have done. We have depended upon a Newtonian concept of time so that we might shed some light on the nature of irreversibility. We take irreversibility to be as fundamental as temporality and time.

NOTES

1. Page 5. For the chronology and descriptive history we are indebted to the scholarly work of Broda (1983), Brush (1967,1976), Dugas (1959), Elkana (1974), Kangro (1976), Klein (1973), and Kuhn (1978). Using this literature we have constructed a narrative about the origins and development of the concept of irreversibility. These authors, of course, bear no responsibility for the particular interpretation offered here.

2. Page 52. Notice the original discussions in Boltzmann (1872) and (1964); the first major defense, Ehrenfest (1959); and some of the recent reviews, meetings, symposia, e.g., ter Haar (1954) (see Appendix 1, pp.331-385); Prigogine (1958) (notice especially the articles beginning on pp.1,8,15,25,132,219 and the discussions beginning on pp.19,135, and later). (Notice that the question of irreversibility is raised repeatedly and passed over.) Cohen (1962): (Notice especially the articles by Nijboer, Cohen, Van Hove, and Van Kampen); Gold (1967) and Garrido (1972) (notice the excellent introduction and review by Rosenfeld and the articles by Farquhar and Wergeland). The Boltzmann Equation, edited by Cohen and Thirring (1973); notice especially the articles by Klein, Cohen, Kac, and Prigogine. See also Landsberg (1970) and de Beauregard (1964).

3. Page 52. See, for example, pp.446-447 of Boltzmann (1964); p.384 of Reichenbach (1956); pp.30,31 of Landau and Lifshitz (1958); Denbigh (1961); Melberg (1961); pp.25-30 of Schlegel (1961); pp.149-186 of Gold (1969); pp.319-326 of Liboff (1969); pp.49,293 of Bastin (1971); pp.9,10 of Garrido (1972); Chapter 8 of Grunbaum (1973); Layzer (1975); Mirman (1975); and less direct statements in many of the other references cited.

4. Page 58. The "reversible" electrode is defined somewhat differently in electrochemistry, involving a mixture of the criteria for reversibility suggested here, and for good reasons.

5. Page 63. See, for example, Eq.(42) of Hollinger and Curtiss (1960). The original equation can be recovered explicitly after time reversal by replacing the integration variable p_2 by $-p_2$ and then changing signs to

restore the original limits for the p_2 integration. The conclusion is not altered by examining the P functions. More generally, sufficiently explicit forms of the Boltzmann equation are reversible while less explicit forms are difficult to classify because of the complex meaning of the symbols that refer to inverse collisions. Notice, for example, the Boltzmann collision integral in Eq.(9.246) of this book. If we replace $\underline{v}_1$ by $-\underline{v}_1$ then we can recover the integral by replacing integration variables $\underline{v}_2$ and $\underline{k}$ by $-\underline{v}_2$ and $-\underline{k}$. The sign change of $\underline{k}$ is required to maintain $(\underline{v}_{21} \cdot \underline{k}) < 0$ which restricts $\underline{k}$ to directions that cause the $\underline{v}_i$ to be precollision velocities. After these changes the result J does not appear to be the same as before with or without a sign change. But if we see the primes represented more explicitly by the S_{12} operator in (9.293) then we realize there are primes in both terms on the right side of (9.246). The primes on the second term do not change the $\underline{v}_i$, which are already precollision velocities. After the changes of sign of the $\underline{v}_i$ and $\underline{k}$, the primes become active in that term and inactive in the other term. Thus J is recovered after imposing the test of invariance under time reversal and its sign has changed. The integral J does change sign under time reversal and hence the Boltzmann equation is invariant. The Boltzmann equation (9.253) and the Liouville equation (9.253) with J replaced by (9.242), do not differ in their degree of temporal irreversibility. The difference is that the Liouville equation is always reliable and never deterministic while the Boltzmann equation is always deterministic but reliable only near recent equilibrium.

6. Page 67. See, for example, Tolman (1938, p.152); Ehrenfest (1959, p.13); Prigogine (1955, pp.14,15); Prigogine (1958, pp.15,16,20,28,65-72,340); Landau and Lifshitz (1958, p.30); Schlegel (1961, p.48); Mehlberg (1961); Cohen (1962, pp.144,174,178); Gold (1963, p.25). Liboff (1969, p.180); Garrido (1972, p.116); Mirman (1975, p.492); and many other references. This interpretation of reversibility is taken for granted in many articles, as for example, the articles in Brittin (1961); see especially the statement by Montroll about the need for statistical games to beat (temporal) reversibility and Chapter 8 of Prigogine (1980). The assumption that the problem of irreversibility is

centered around temporal irreversibility is also explicit in most modern texts which deal with irreversible processes. The question has been treated much more carefully in many articles where the issue was in the air but not trivialized. See, for example, Denbigh (1961), de Beauregard (1964), Grad (1967), and Landsberg (1972). The article by Landsberg is very helpful here, and his Appendices A and B should not be passed over.

7. Page 69. The early discussions of interest here can be found in Mehlberg (1961). Those and later discussions are fully referenced by Grübaum (1973).

8. Page 72. From a very good point of view, which may be naive against future standards, we could say that the reproducibility of macroscopic phenomena is incredible. That puzzle is bigger than any other one we could consider concerning the emergence of order from disorder. The question is not why statistical mechanics works in the sense of giving unique answers to predictive questions, but why nature with all of its potentialities for diverse evolutions, chooses to obey this dumb tool of man. The answer to the question is known at equilibrium. The overwhelming majority of all those diverse evolutions look alike at the fluid level when the fluid is at equilibrium. Thus the question that plagues us is why the fluid <u>evolutions</u> are <u>alike</u> when the fluid is <u>not</u> at equilibrium. That is the uppermost question in our thoughts as we develop the arguments presented here. Can we say that all the possible diverse evolutions look alike at the macroscopic level? Certainly not, or so it would seem. Where is the origin of reproducibility?

9. Page 73. This, according to Gibbs, is the most fundamental definition of entropy, more fundamental than

$$S = k \ln W$$

which is often ascribed to Boltzmann. The point is that Boltzmann's formula works when W is a good estimate of Ω, but to be on the safe side, we need Ω, not W. It is generally taken for granted that Ω and W are the same thing, even by Tolman (1938), Hill (1962) and Morse (1964), and rightly so because they are preoccupied with equilibrium. Away from equilibrium we can continue to equate Ω and W, but only if the W is for the <u>most</u>

probable distribution, which may be different from the actual distribution (another W). All authors are willing to start with Ω in the definition of entropy and to replace S by the most probable W, but few are concerned about the difference between the most probable W and the evolved W away from equilibrium. When in doubt, go back to Ω.

10. Page 74. It would be better for some purposes to refer to the momentum p and Δp rather than v and Δv, and to divide $\Delta x \Delta p$ by Planck's constant h. In quantum mechanics the "number of states" in an area $\Delta x \Delta p$ is that area divided by h.

11. Page 78. See Boltzmann (1964, especially pp.441-448). See also Jaynes (1965).

12. Page 79. For descriptions of the paradoxes, see, for example, Ehrenfest (1959, pp.14-16); Liboff (1969, p.182); ter Haar (1954). For a recent disclaimer, see Rothstein (1974). For a deeper analysis with historical perspective, see Brush (1967) and Elkana (1974).

13. Page 84. See Chapman and Cowling (1952, Chap.7). The first order correction $f^{(1)}$ used in the calculation of transport coefficients represents the difference between the evolved f and the local equilibrium or most probable f.

14. Page 85. This is our first and simplest illustration of a general theorem which is known as the "H-Theorem" of statistical mechanics. We will encounter more complex illustrations below. Actually, in spite of its physical simplicity, the H-theorem here is "better" than the ones described below because it really does deal with the actual entropy of the fluid considered. In later illustrations of the H-theorem we will have a problem identifying the fluid for which the theorem holds.

15. Page 89. See for example, Prigogine (1958, p.65); Cohen (1962, p.182); Bastin (1971, p.29); Prigogine (1962, pp.8,228). The tendency to anthropomorphize, or at least psychologize the losses associated with irreversibility has become so commonplace that it was necessary for Landsberg (1970) to specifically rule out psychological and biological mysticisms. We were warned against this tendency by the fathers of Information Theory (Lewis, 1930), who thought they might be

misunderstood. There may be a need for discussions about information, causality, will, determinism, etc., and it may be useful to introduce time dissymmetries, but it does not follow from this that time dissymmetries of the same kind arise in the evolutions of isolated mechanical systems.

16. Page 91. It is assumed that the set of accessible states is "invariant" in the sense that it evolves into itself and is "indecomposable" in the sense that the set cannot be subdivided into subsets of similar measure that are themselves invariant.

17. Page 92. See, for example, Part I of Prigogine (1958); the article by Cohen (1962); Chapters 2 and 3 of Liboff (1969); Hollinger and Curtiss (1960,1962); Bogoliubov (1961); Chapter 6 of Brush (1972).

18. Page 93. See, for example, pp.16,20,21,28,39,70,219-228 of Prigogine (1958); pp.18,81 of Garrido (1972); Van Hove (1955,1957). The interest in large-time perturbation theory was introduced into statistical mechanics mainly by the schools of Prigogine and Van Hove. For perspectives in this area, see the articles by Zwanzig and Montroll in the volume edited by Brittin (1961).

19. Page 93. The search for "asymptotic" formulas in the sense of large N is explicit in the references of Note 18 and many other references that could be listed here. See, for example, pp.48,49 of Schlegel (1961). We are reminded of the search for asymptotic formulas in equilibrium statistical mechanics, where the motivation is completely justified in terms of the "law of large numbers" of probability and the smallness of fluctuations at the macroscopic level. In fact, there must be a similar asymptotic character away from equilibrium, probably inherited from recent equilibrium. But the motivation behind the restrictions to asymptotic formulas in the perturbation theories is quite different. The idea is that irreversibility itself is an asymptotic thing and cannot be identified by descriptions of finite systems. From our point of view, irreversibility can emerge in systems of any size depending on the relation between typical lengths of equilibrium plateaus and times between interruptions due to changes in the environment.

20. Page 93. See p.28 of Prigogine (1958); p.185 of Cohen (1962); pp.4,5 of Prigogine (1962); pp.416,420 of Balescu (1975). The tendency to attribute irreversibility and dissipation to interactions is quite understandable in view of familiar examples, e.g., the decay of systems that have friction among the moving parts. But the dissipation is just as severe and much faster in the expansion of a collisionless gas.

21. Page 94. See p.1 of Prigogine (1958); pp.144,178 of Cohen (1962); p.17 of Garrido (1972); pp.326-340, 367-370 of Liboff (1969); pp.185,186 of Bastin (1971); pp.17-23 of Farquhar (1964); pp.83,84 of de Groot and Mazur (1962).

22. Page 94. See pp.39,219 of Prigogine (1958); pp.18-26 of Garrido (1972); p.8 of Prigogine (1962); Chapter 14 of Balescu (1975).

23. Page 95. See pp.132-135 of Prigogine (1958); pp.7-20 of Gold (1967); pp.10,11 of Garrido (1972); pp.4,36 of Prigogine (1962); pp.48,120,121 of Fugita (1966); p.467 of de Beauregard (1970).

24. Page 96. See p.347 of Prigogine (1958); p.17 of Garrido (1972); p.31 of Landau and Lifshitz (1958); pp.200-203 of Farquhar (1964).

25. Page 103. See Brittin (1959, p.222). The counterpoint is expressed very well by Gal-Or (1976, p.635). The argument by Montroll is useful in many ways because it makes more explicit some of the tacit assumptions involved in asymptotic theories. For example, Montroll introduces explicitly the distinction between "apparent" and "true" irreversibility. True irreversibility requires infinite N to thoroughly destroy the Poincaré periodicity. From our point of view, true irreversibility requires no such infinities, nor continuous spectra of operators nor large times, etc. It merely requires equilibrium plateaus that are longer than the times between substantial interactions with a changing environment. In other words, the only connection between irreversibility and the Poincaré periodicity is that the periods have to be long compared to times between "interruptions."

26. Page 128. The signs of off-diagonal elements of the pressure tensor have been derived with difficulty (finding the most convenient convention) in fluid dynamics. The choices become obvious in the molecular interpretation considered in Chapter 9. In the molecular interpretation, shearing forces are momentum fluxes and it is always easy to decide whether a flux is in the positive or negative direction.

27. Page 147. The "principle of microscopic reversibility" was introduced by Tolman (1924) and later introduced into quantum theory by Mitchell and Zemansky. The principle is very useful and is needed to show that at macroscopic equilibrium each molecular process is separately equilibrated ("principle of detailed balance"). Tolman's motivation for the choice of words, which he later questioned (1938, p.163), was to emphasize a distinction between reverse and inverse collisions. Then the ideas about detailed balance became a part of the derivation of the reciprocal relations. See, for example, Casimer (1945) and the article by Mazur in Cohen (1961). For an explicit reference to the role of temporal reversibility at the molecular level, see the article by de Groot and Mazur (Prigogine, 1958). The connection between temporal reversibility and the reciprocal relations received more and more emphasis and is now always mentioned in texts on irreversible thermodynamics. The role of equilibrium in the cause of reciprocity is explicit in derivations (Cox, 1950) but not in other contexts. Thus the term "microscopic reversibility" has taken on a suggestive role that was never intended by Tolman. It is seen as a manifestation of reversibility at the molecular level to be held against irreversibility at the fluid level. In fact, the temporal reversibility at the microscopic level holds whether or not the system is inertially reversible at the fluid level. The special requirement for the reciprocal relations is reversibility at the macroscopic fluid level.

28. Page 304. Time here is taken to be the concept time. Time as a concept, of course, has a complex history but in contemporary science it is the linear, uniform, homogeneous, continuum, represented by "t" in physics and mechanics. Despite debates about temporal anisotropy and time's arrow, quantization of time, and relativization of time, the equations of normal science do not reflect these variations. Formally, the concept of time used in science has not changed, since Newton. Temporality is used here to indicate time as lived or experienced. In other words, time as phenomenon. Relevant philosophical treatments can be found in Bergson, James, Whitehead, Husserl, Heidegger, and Merleau-Ponty.

It might be helpful to characterize the difference between time and temporality as one of structure and significance. At one end of a spectrum there is the purely formal structure of Newtonian time, the "t" of classical mechanics and at the other end, "temporality" which is inseparable from subjectivity and significance.

By making this distinction we immediately surrender the security of our naive approach!

29. Page 310. For a deeper analysis, see Denbigh, Three Concepts of Time (1981). Notice his comparison of A-theorists and B-theorists.

REFERENCES

Adams, H.: 1919, "The Rule of Phase Applied to History," in The Degradation of Democratic Dogma, The Macmillan Co., New York

Alder, B.J. and Wainwright, T.: 1958, "Molecular Dynamics by Electronic Computers," in I. Prigogine (ed.), Transport Processes in Statistical Mechanics, Interscience Publishers, New York.

Atkins, P.N.: 1980, Physical Chemistry, W.H. Freeman, San Francisco.

Balescu, R.: 1963, Statistical Mechanics of Changed Particles, Interscience Publishers, New York.

Balescu, R.: 1975, Equilibrium and Nonequilibrium Statistical Mechanics, John Wiley and Sons, New York.

Bastin, T.: 1971, Quantum Theory and Beyond, Essays and Discussions Arising from a Colloquium, Cambridge University Press, London.

Bendixson, I.: 1901, "Sur les courbes définies par des équations différentielles," Acta Math, 24, 1-88.

Bent, H.: 1965, The Second Law, Oxford University Press, New York.

Biolsi, L. and Hollinger, H.: 1974, "The Withdrawal from Equilibrium of a Collisionless Gas," in Rarefied Gas Dynamics, Academic Press, New York.

Bogoliubov, N.: 1946, "Problemy Dinamicheskoi Teorii v Statistichestoi Fizike," Technical Press, Moscow.

Bogoliubov, N.: 1961, "Problems of Dynamical Theory in Statistical Physics," in J. de Boer and G. Uhlenbeck (eds.), Studies in Statistical Mechanics, Vol.I, Interscience Publishers, New York.

Boltzmann, L.: 1866, "Über die mechanische Bedeutung der zweiten Hauptsatzes der Wärmetheorie," Wien. Ber. 53, 195.

Boltzmann, L.: 1964, Lectures on Gas Theory, translation of 1896-98 volumes published by J.A. Barth, Leipzig, S. Brush (tr.), University of California Press, Berkeley.

Boltzmann, L.: 1872, "Weitere Studien über die Wärmegleichgewicht unta Gasmolekülen, Wien. Ber., II, 66, 275-370.

Born, M.: 1948, "Die Quantenmechanik und zweite Hauptsatz der Thermodynamik," Annalen der Physik, 3, 197.

Born, M.: 1949, Natural Philosophy of Cause and Chance, Clarendon Press, Oxford.

Born, M. and Green, H.: 1949, A General Kinetic Theory of Liquids, Cambridge University Press, New York.

Brandon, S.: 1972, "The Deification of Time," in J. Fraser (ed.), The Study of Time, Springer-Verlag, New York.

Brandon, S.: 1965, History, Time, and Deity, Barnes and Noble, New York.

Brittin, W.E. (ed.): 1959, Lectures in Theoretical Physics, Vol.3, Interscience Publishers, New York.

Broda, E.: 1973, Philosophical Biography of L. Boltzmann, in E. Cohen and W. Thirring (eds.), The Boltzmann Equation, Springer-Verlag, New York.

Broda, E.: 1983, Ludwig Boltzmann, Ox Bow Press, Woodbridge, Connecticut.

Brout, R. and Prigogine, I.: 1956, Physica, 22, 621.

Brush, S.G.: 1965,66,72, Kinetic Theory, Volumes 1, 2, 3, Pergamon Press, Oxford.

Brush, S.: 1967, "Foundations of Statistical Mechanics 1845-1915," Archives for History of Exact Sciences, 4, 145.

Brush, S.G.: 1976, The Kind of Motion We Call Heat, North Holland Publishing Co., Amsterdam.

Burbury, S.H.: 1899, A Treatise on the Kinetic Theory of Gases, Cambridge University Press, London.

Burnett, D.: 1935, Proc. London Math. Soc., 40, 382.

Carmody, W.: 1960, "Dynamic Equilibrium," J. Chem. Ed., 37, 312.

Chapman, S.: 1916, Phil. Trans. R. Soc. A., 216, 279.

Chapman, S. and Cowling, T.: 1939, 1952, 1969, Mathematical Theory of Nonuniform Gases, Cambridge University Press, London.

Choh, S.T.: 1958, "The Kinetic Theory of Dense Gases," Dissertation, University of Michigan.

Clark, P.: 1974, Atomism vs Thermodynamics, read before the British Society for the Philosophy of Science on April 29, 1974.

Clausius, R.: 1852, Proc. R.S., 3, 139, Edinburgh.

Clausius, R.: 1865, "On Several Convenient Forms of the Fundamental Equations of the Mechanical Theory of Heat," Pogg. Ann., 125.

Coddington, E. and Levinson, N.: 1955, Theory of Ordinary Differential Equations, McGraw-Hill Publishing Co., New York.

Cohen, E.G.D. (ed. and author): 1962, "Fundamental Problems in Statistical Mechanics," Proceedings of the NUFFIC International Summer Course, Interscience Publishers, New York.

Cox, R.: 1950, "The Statistical Method of Gibbs in Irreversible Change," Rev. Modern Phys., 22, 238.

Curtiss, C.: 1948, "The Equilibrium Assumption in the Theory of Absolute Reaction Rates," Report CM-476, University of Wisconsin.

Daub, E.: 1969, "Probability and Thermodynamics: The Reduction of the Second Law," Isis, 60, 318.

Davies, P.C.W.: 1974, The Physics of Time Asymmetry, University of California Press, Berkeley.

Davies, P.C.W.: 1977, Space and Time, Cambridge University Press, Cambridge.

de Beauregard, O.: 1964, "Irreversibility Problems," Proceedings of the International Congress for Logic, Methodology, and Philosophy of Science, North Holland Publishing Company, Amsterdam.

de Beauregard, O.: 1970, "Is There a Paradox in the Theory of Time Anisotropy?", A Critical Review of Thermodynamics, Mono Book Company, Baltimore.

de Broglie, L., Lochak, G., Beswick, J. and Vassalo-Pereira, J.: 1970, "Present, Predicted and Hidden Probabilities," Foundations of Physics, 6, 3.

de Groot, S. and Mazur, P.: 1962, Nonequilibrium Thermodynamics, North Holland Publishing Co., Amsterdam.

Denbigh, K.: 1971, Chemical Equilibrium, Cambridge University Press, London.

Denbigh, K.: 1978, "In Defense of the Direction of Time," in J. Fraser (ed.), The Study of Time, Springer-Verlag, New York.

Denbigh, K.: 1981, Three Concepts of Time, Springer-Verlag, New York.

Dugas, R.: 1959, "La théorie physique au sens de Boltzmann," Editions du Griffon, Neuchâtel.

Ehrenfests, P. and T.: 1959, "The Conceptual Foundations of the Statistical Approach in Mechanics," (tr. of 1911 article in Encyklopadie der mathematischen Wissenchaftsen, Teubner, Leipzig), Cornell University Press, Ithaca.

Einstein, E.: 1922, "Zur Theorie des Radiometers," Annalen der Physik, 69, 241.

Elkana, Y.: 1974, "Boltzmann's Scientific Research Program and its Alternatives," The Interaction Between Science and Philosophy, Humanities Press. Atlantic Highlands, New Jersey.

Enskog, D.: 1917, "Kinetische Theorie der Vorgänge in mässig verdünnten Gasen," Ph.D. Dissertation at Uppsala and in Archiv für Mathematik Astronomi, och Fysik, 16, 1.

Enskog, D.: 1922, "Kinetic Theory of Thermal Conduction, Viscosity, and Self-Diffusion in Certain Dense Gases and Liquids," Kgl. Svenska Vetenskass Akad. Handl., 63, No.4.

Eyring, H.: 1935, J. Chem. Phys., 3, 107.

Eyring, H.: 1936, J. Chem. Phys., 4, 283.

Farquhar, I.: 1964, Ergodic Theory in Statistical Mechanics, Interscience Publishers, New York.

Fast, J.: 1962, Entropy, McGraw-Hill Book Company, New York.

Fetter, A.L. and Walecka, J.D.: 1971, Quantum Theory of Many-Particle Systems, McGraw-Hill Book Company, New York.

Fowler, R. and Guggenheim, E.: 1939, Statistical Thermodynamics, Cambridge University Press, London.

Fraser, J.T., Haber, F.C. and Miller, G.H.: 1972, The Study of Time, Springer-Verlag, New York.

Fugita, S.: 1966, Introduction to Nonequilibrium Statistical Mechanics, W.B. Saunders and Co., Philadelphia.

Gal-Or, B.: 1974, Modern Developments in Thermodynamics, John Wiley, New York.

Gal-Or, B.: 1976, "Cosmological Origin of Irreversibility, Time and Time Anisotropies I and II," Foundations of Physics, 6, Nos.4 and 6.

Gal-Or, B.: 1981, Cosmology, Physics, and Philosophy, Springer-Verlag, New York.

Garrido, L.M. (Director): 1972, "Irreversibility in the Many Body Problem," Sitges International School of Physics, Plenum Press, New York.

Gibbs, T.W.: 1902, Elementary Principles in Statistical Mechanics, Yale University Press, New Haven.

Glasstone, S., Laidler, K. and Eyring, H.: 1941, The Theory of Rate Processes, McGraw-Hill Book Co., New York.

Gold, T.: 1962, "The Arrow of Time," Amer. J. Phys., 30, 403.

Gold, T. (ed.): 1967, The Nature of Time, A Meeting at Cornell University, Cornell University Press, Ithaca.

Grad, H.: 1961, "The Many Faces of Entropy," Communications of Pure and Applied Mathematics, 14, 323.

Grad, H.: 1967, "Levels of Description in Statistical Mechanics and Thermodynamics," Delaware Seminar on the Foundations of Physics, Vol.I, Springer-Verlag, New York.

Grecos, A. and Prigogine, I.: 1972, "Kinetic and Ergodic Properties of Quantum Systems," Physica, 59.

Green, H.: 1952, Molecular Theory of Fluids, Interscience Publishers, New York.

Green, M.S.: 1956, "Boltzmann Equation from the Statistical Mechanical Point of View," J. Chem. Phys., 25, 836.

Green, M.S.: 1961, "Some Applications of the Generating Functional of the Molecular Distribution Functions," in W. Brittin (ed.), Lectures in Theoretical Physics, Vol.III, Interscience Publishers, New York.

Grünbaum, A.: 1973, "Philosophical Problems of Space and Time," in Volume 12 of M. Wartofsky (ed.), Boston Studies in the Philosophy of Science, D. Reidel Publishing Co., Boston.

Hiebert, E.: 1966, "The Uses and Abuses of Thermodynamics in Religion," Daedalus, 95.

Hilbert, D.: 1912, "Grundzüge einer allgemeinen Theorie der linearen Integralgleichungen," Math. Annalen, 72, 567.

Hill, T.: 1960, Introduction to Statistical Thermodynamics, Addison Wesley Publishing Company, Reading.

Hirschfelder, J.O., Curtiss, C.F. and Bird, R.B.: 1954, Molecular Theory of Gases and Liquids, John Wiley, New York.

Hollinger, H.B. and Curtiss, C.F.: 1960, "Kinetic Theory of Dense Gases," J. Chem. Phys., 33, 1386-1402.

Hollinger, H.: 1962, "Molecular Chaos and the Boltzmann Equation," J. Chem. Phys., 36, 3208.

Hollinger, H.: 1972, "Equilibrium," The Science Teacher, 39, No.7.

Jammer, M.: 1974, The Philosophy of Quantum Mechanics, John Wiley and Sons, New York.

Jancel, R.: 1969, Foundations of Classical and Quantum Mechanics, Pergamon Press, Paris.

Jaynes, E.: 1965, "Gibbs vs Boltzmann Entropies," American Journal of Physics, 33, 191.

Jaynes, E.: 1967, "Foundations of Probability Theory and Statistical Mechanics," Delaware Seminar in Foundations of Physics, Springer-Verlag, New York

Jeans, J.H.: 1925,1954, The Dynamical Theory of Gases, Cambridge University Press (1925) and Dover (1954), New York.

Kac, M.: 1954,1959, Probability and Related Topics in Physical Sciences, Berkeley Symposium (1954) and Interscience Publishers (1959), New York.

Kangro, H.: 1976, History of Planck's Radiation Law, Taylor and Francis, London.

Khinchin, A.: 1949, Mathematical Foundations of Statistical Mechanics, Dover, New York.

Kirkwood, J.G.: 1946, J. Chem. Phys., 14, 180, and 1947, 15, 72.

Kirkwood, J.G.: 1958, "The Statistical Mechanical Basis of the Boltzmann Equation," in I. Prigogine (ed.), Transport Processes in Statistical Mechanics, Interscience Publishers, New York.

Klein, M.J.: 1973, "The Development of Boltzmann's Statistical Ideas," in E. Cohen and W. Thirring (eds.), The Boltzmann Equation, Springer-Verlag, New York.

Klein, M.J.: 1974, "Boltzmann, Monocycles, and Mechanical Explanation," in Philosophical Foundations of Science, 155-175, D. Reidel Publishing Co., Dordrecht.

Koneig, F.: 1959, "The History of the Second Law of Thermodynamics," Men and Moments in the History of Science, H. Evans (ed.), University of Washington Press, Seattle.

Kuhn, T.S.: 1978, Blackbody Theory and the Quantum Discontinuity in 1894-1912, Oxford University Press, New York.

Landau, L. and Lifshitz, E.: 1958, Statistical Physics, Addison-Wesley Publishing Co., Reading.

Lande, A.: 1953, Scientific Papers Presented to Max Born, Oliver and Boyd, Edinburgh.

Landsberg, P.T.: 1970, "Time in Statistical Physics and Special Relativity," Stadium Generale, 23, 1108-1158, Springer-Verlag, New York.

Layzer, D.: 1975, "The Arrow of Time," Scientific American, December.

Lewis, G.: 1930, "The Symmetry of Time in Physics," Science, 71, 569.

Liboff, R.: 1969, Introduction to the Theory of Kinetic Equations, John Wiley and Sons, New York.

Maxwell, J.C.: 1867, Collected Papers, 2, 42; Phil. Trans. R. Soc., 157.

Mehlberg, H.: 1961, "Physical Laws and Time's Arrow," Current Issues in the Philosophy of Science, Holt, Rinehart, and Winston, New York.

Mirarchi, L.: 1967, The Quantum Mechanical Basis of Chemical Kinetics, Ph.D. Dissertation, Rensselaer Polytechnic Institute, Troy.

Mirman, R.: 1975, "The Direction of Time," Foundations of Physics, 4, 491.

Montroll, E.: 1961, "Nonequilibrium Statistical Mechanics," in W. Brittin (ed.), Lectures in Theoretical Physics III, Interscience Publishers, New York.

Moore, W.: 1972, Physical Chemistry (4th ed.), Prentice Hall, Inc., Englewood Cliffs.

Morowitz, H.J.: 1979, Energy Flow in Biology, Ox Bow Press, Woodbridge, Connecticut.

Morrissey, B.: 1970, Irreversible Processes in Rarefied Gases, Ph.D. Dissertation, Rensselaer Polytechnic Institute, Troy.

Morrissey, B.: 1975, "Microscopic Reversibility and Detailed Balance," J. Chem. Ed., 52, 296.

Morse, P.: 1964, Thermal Physics, W.A. Benamin, New York.

Muriel, A. and Dresden, M.: 1969, Projective Techniques in Nonequilibrium Statistical Mechanics, North Holland, Amsterdam.

Olson, R.: 1971, Science As Metaphor, Wadsworth Publishing Co., Inc., Belmont, California.

Onsager, L.: 1931, Phys. Rev., 37, 405, and Phys. Rev., 38, 2265.

Petersen, R.: 1984, Binary Diffusion of Gases, Improvements of the Mean Free Path Theory, Ph.D. Dissertation, Rensselaer Polytechnic Institute, Troy.

Poincaré, H.: 1892, "Sur les courbes définies par une équation différentielle," Oeurres, 1, Paris.

Prigogine, I.: 1955, Thermodynamics of Irreversible Processes, Interscience Publishers, New York.

Prigogine, I. (ed.): 1958, Proceedings of the International Symposium on Transport Processes in Statistical Mechanics (Brussels, 1956), Interscience Publishers, New York.

Prigogine, I.: 1962, Nonequilibrium Statistical Mechanics, Interscience Publishers, New York.

Prigogine, I.: 1980, From Being to Becoming, W.H. Freeman, San Francisco.

Prigogine, I. and George, C.: 1983, Natl. Acad. Sci. U.S.A., 80, 4590.

Reichenbach, H.: 1956, The Direction of Time, University of California Press, Berkeley.

Reichenbach, H.: 1958, The Philosophy of Space and Time, Dover, New York.

Rice, S.A., Kirkwood, J.G., Ross, J. and Zwanzig, R.W.: 1959, J. Chem. Phys., 31, 575.

Rothstein, J.: 1974, "Loschmidt's and Zermelo's Paradoxes Do Not Exist," Foundations of Physics, 4, 83.

Schiff, L.I.: 1955, Quantum Mechanics, McGraw-Hill Book Co., New York.

Schlegel, R.: 1961, Time and the Physical World, Michigan State University Press, East Lansing.

Schrödinger, E.: 1944, What is Life?, Cambridge University Press, London.

Sears, F.: 1956, Thermodynamics, Kinetic Theory of Gases, and Statistical Mechanics (2nd ed.), Addison-Wesley Publishing Co., Reading.

Shannon, C.E. and Weaver, W.: 1949, The Mathematical Theory of Communication, University of Illinois Press, Urbana.

Sklar, L.: 1973, "Statistical Explanation and Ergodic Theory," Philosophy of Science, 40, 194.

Strauss, M.: 1972, "Max Planck and the Rise of Quantum Theory," Modern Physics and Its Philosophy, D. Reidel Publishing Co., Dordrecht, Holland.

Szilard, L.: 1929, Z. Physik., 53, 840.

Tait, P.G.: 1886, Trans. R. Soc. Edinb., 33, 74.

ter Haar, D.: 1954, Elements of Statistical Mechanics, Rinehart and Company, New York.

Tolman, R.: 1938, The Principles of Statistical Mechanics, Oxford University Press, London.

Uhlenbeck, G.E.: 1957, The Boltzmann Equation, University of Michigan.

Van Hove, L.: 1955, Physica, 21, 517.

Van Hove, L.: 1957, Physica, 23, 441.

Wannier, G.: 1966, Statistical Physics, John Wiley and Sons, New York.

Watanabe, S.: 1951, "Contribution to Louis de Broglie," Revue de Métaphysique et de Morale, Albin Michel, Paris.

Watanabe, S.: 1955, "Symmetry of Physical Laws," Rev. Mod. Phys., 27.

Webster, A.G.: 1955, Partial Differential Equations of Mathematical Physics, Dover Publications, Inc., New York.

Weiner, N.: 1948, Cybernetics, New York.

Yang, H. and Lees, L.: 1960, "Raleigh Problem at Low Reynolds Number According to Kinetic Theory," Proceedings of the First International Symposium on Rarefied Gas Dynamics, Pergamon Press, New York.

Yvon, J.: 1935, Actualities Scientifiques et Industrielles, Herman, Paris.

Zenzen, M.: 1977, "Popper, Grünbaum and de facto Irreversibility," British Journal for Philosophy of Science, 28, 313.

Zwansig, R.: 1961, "Statistical Mechanics of Irreversibility," in W.E. Brittin (ed.), Lectures in Theoretical Physics, III, Interscience Publishers, New York.

NAME INDEX

SUBJECT INDEX

GPSR Compliance

The European Union's (EU) General Product Safety Regulation (GPSR) is a set of rules that requires consumer products to be safe and our obligations to ensure this.

If you have any concerns about our products, you can contact us on ProductSafety@springernature.com

In case Publisher is established outside the EU, the EU authorized representative is:

Springer Nature Customer Service Center GmbH
Europaplatz 3
69115 Heidelberg, Germany

Batch number: 10761599

Printed by Printforce, the Netherlands